Wilhelm Heinrich
Schu
..
ssler, Louis H. Tafel

An Abridged Therapy
Manual for the Biochemical Treatment of Disease

ISBN/EAN: 9783337407926

Printed in Europe, USA, Canada, Australia, Japan

Cover: Foto ©berggeist007 / pixelio.de

More available books at **www.hansebooks.com**

AN ABRIDGED THERAPY.

MANUAL

FOR THE

Biochemical Treatment of Disease.

BY

DR. MED. SCHUESSLER,

OF OLDENBURG.

Twenty-Fifth Edition, in Part Rewritten.

TRANSLATED BY

PROF. LOUIS H. TAFEL.

PHILADELPHIA:
BOERICKE & TAFEL.
1898.

PUBLISHER'S PREFACE.

It was our intention, and Dr. Schüssler's wish, that nothing should go into this translation of the *Abgekürzte Therapie* save what is to be found in the original. At his request we waited until the 25th edition of the German was published, which differs materially from preceding editions, and have now made our translation in accordance with his desire to have what he wrote on the subject of biochemistry, nothing more and nothing less, given to the English-speaking world; something, he claimed, that had never been done before. The author's death, however, following immediately on the completion of his work has caused us to so far depart from his request as to insert as a preface, a short account of his life and to give a *fac simile* of one of his letters. Otherwise the book is a strictly literal translation of the 25th edition of his work, and the only authorized translation of this work on Biochemistry published.

BOERICKE & TAFEL.

DR. SCHUESSLER.*

OBITUARY.

Our honored master, the founder of Biochemistry, is no more among the living. The wish of his many adherents at home and in the wide, wide world, that he might be permitted to labor for yet many years with his accustomed bodily vigor and mental energy in advancing the work of his life, has not been fulfilled. Dr. Wilhelm Heinrich Schüssler died on March 30th, in consequence of an apoplectic stroke. Up to the morning of March 14th he had felt in good health, but then he had a stroke; he quickly recovered, however, so that he was able to finish on the afternoon of the following day the last proof of the last sheet of the 25th edition of his Abridged Therapy. This was fated to be his last work, for the improvement did not continue, perhaps the patient himself being to blame, as he would in no way spare him-

---

*Translated from *Mitth. ueber Biochemie*, May, 1898.

self. Soon his condition became so much aggravated that no one could doubt his approaching end, of which he himself was conscious, and the approach of which he saw coming with the greatest tranquillity. Having been unconscious for several days he expired on the evening of March 30th.

In Dr. Schüssler not only his friends and adherents, but all mankind have lost one of the best of men. The value to mankind of the deceased as a physician and teacher, posterity will be able to appreciate better than the present time, although even now many signs of appreciation are manifested. Dr. Schüssler was not only a learned and important man in his own domain, that of medicine, but he was also eminent in other domains of knowledge. He had a peculiar genius for the study of foreign tongues, and was a perfect master not only of Latin and Greek, but also of French, Italian, Spanish and English. His love for comparative philology had induced him also to study Sanskrit.

Through his system of Therapy, Dr. Schüssler has become known throughout the civilized world, and from all parts of the world patients came to him to get his medical advice. But all his great successes, much as he enjoyed their recognition, did not make him proud; he always remained the plain and simple man he had been from his youth. Although he lived in his own large residence, in one of the finest streets of Oldenburg, his furniture was no richer than that of many a citizen in moderate circumstances. Making money was always a very subordinate matter with Dr. Schüssler; the main point with him was always the cure of his patients and the development of his Therapy. His fees during the whole of his medical life were always low, and many families, which he for years had been treating gratuitously, will miss him bitterly. If he, nevertheless, acquired a comparatively large property, this is to be ascribed to his extensive medical practice and his very limited personal requirements. That he had a fellow-feeling also for the less wealthy among his fellow-citizens is shown by the particulars of his last will and testament.

A prominent feature of Schüssler's character was his straightforwardness, which sometimes, especially when something was imputed to him which he could not reconcile with his views, passed over into roughness, and this without regarding whether his opponents were men of distinction or common people. Free from the fear of man, he went his way without minding whether he gave offence on the right hand or on the left; and full of conviction of his principles, he defended his cause against all. He was a man of character in every way. Even his opponents, in so far as their judgment is unbiased, agree with his friends and adherents in the unanimous recognition of his worth. Those who intimately knew and understood Dr. Schüssler, the few whom he deemed fully worthy of his confidence, cannot do otherwise but say with Hamlet:

> "He was a man, take him for all in all,
> I shall not look upon his like again."

But little is known as to the life and development of the creator and founder of biochemistry. Hardly anything touching it is found in the papers he has left behind him, and there are no near relatives living—he was unmarried—who might give us information. The repeated requests of his friends to write an autobiography, he had always put aside with the utmost decision; for while he was fully convinced of the importance and scientific exactness of the therapy created by him, he was reticent and modest in everything touching himself personally.

Schüssler was born on August 21, 1821, in Zwischenalm, in the Grand duchy of Oldenburg, and there passed his childhood. He used his youth and early manhood to acquire an extensive knowledge in various domains of human knowledge, especially in philology. In this he was supported by rare talents, and he could soon successfully act as private teacher. Thus he acquired the scientific basis for his later studies in the universities. Only at a mature age Schüssler could carry out his long-desired wish of entering a university. He studied in Paris, Berlin and Giessen. And in the latter place, after a study of five terms, he acquired his diploma. Then he studied three more terms in Prague.

Besides his studies in the general medical branches, Schüssler also took up the study of Homœopathy, in which he later on distinguished himself.

After the newly-created doctor had also passed the examination at the "Gymnasium" in Oldenburg, and before the *Collegium Medicum* there, the medical examination required by the State, on August 14th, 1857, he received the license—then still required—of settling as physician in Oldenburg. From the first he practised according to the homœopathic curative system.

By many successful cures Dr. Schüssler acquired a great name in the whole country as a homœopathic physician; but far beyond his native land he became known as the founder of a new curative method, that of biochemistry.

Incited by the study of the works of Moleschott and Virchow, he began about the year 1872 to introduce the inorganic substances contained in the blood and the tissues and which there act as the natural means of function, into his practice as medical remedies.

As may appear from the Preface to his fundamental work on Biochemistry, the "Abridged Therapy," he was induced to found his biochemic therapy by the following words of Moleschott in his "Circle of Life" (*Kreislauf des Lebens*):

"The formation of the organs and their ability to live, are dependent on the necessary quantity of the inorganic constituents. Founded on this it is, that the proper valuation of the relation of the inorganic substances to the various parts of the body, a valuation which neither contemptuously disregards other momenta nor is full of extravagant hopes, promises a glorious future both to medicine and to agriculture. It can no more be doubted in the face of the facts bearing upon it, that the substances which remain behind after combustion, the so-called ashy constituents, are as essential constituents of the formative basis of the tissues, and contribute to determining their species just as much as those substances which are volatilized at combustion. Without a basis that furnishes gelatine there are no bones, and just as little can bone be formed without phosphate of lime, or gristle without the cartilage-salts, or blood without iron, or saliva without calcium chlorate.

"Man is created from air and earth. The activity of plants called him into being. The corpse is disintegrated into air and ashes, in order that it may unfold new powers in new forms through the vegetable kingdom."

This new therapy has become known throughout the world, and there is at this day probably no country in which there are not adherents of biochemistry, and physicians who put it into practice. In the town of Oldenburg, the birth-place of the new doctrine, there are at present five practising physicians active as representatives of biochemistry, and in a Memorial Address, dedicated to their departed teacher, they proudly call themselves his pupils.

The Abridged Therapy has been widely distributed, and translated into several tongues. As far as is known there are two translations into English, two into Spanish and one into French. A third translation into English was being made while the author was still living and entirely in agreement with his wishes, *i. e.*, without any addition at the hand of the translator, and this will be published presently in Philadelphia by Messrs. Boericke & Tafel.

The 25th edition of the Therapy was published shortly before the death of the author, and he still lived to have the pleasure of distributing a number of copies of the same among his colleagues and friends.

The long-cherished hope of his adherents that this new edition might become a jubilee-edition has not, we are sorry to say, been realized; for sadness fills their hearts instead of joy, mourning for their teacher, who died much too soon for humanity; and the 25th edition will ever remind them of how much they have lost.

The body of Dr. Schüssler, accompanied by a numerous mourning procession, was carried to the grave on Tuesday, April 5th, on a sunny, glorious spring morning. The atoms which had been conjoined together in this great man, not merely for joy and for grief, but still more for the fulfillment of high duties, have been restored to mother earth. But the labors of his spirit have not been in vain, and the most distant generations, we fervently believe, will bless the name of Schüssler and his work, Biochemistry.

> "The traces of his earthly life
> Even ages shall not wipe away."

M.

PREFACE.

Dr. Moleschott, Professor of Physiology in the University of Rome, says in his work *Kreislauf des Lebens* (The Cycle of Life) :

" The structure and vitality of organs are conditioned by the necessary amounts of inorganic constituents. It is owing to this fact that the proper estimation of the relation of the inorganic substances to the various parts of the body, an estimation which neither proudly disdains other momenta nor indulges in extravagant hopes for itself, promises to *Agriculture* and to *Medicine* a brilliant future. In view of all the facts bearing on the case, it can no more be controverted that the substances remaining after combustion—the so-called ashy constituents—belong

2

just as essentially to the internal constitution, and thereby to the basis of the tissues which gives to them their form and determines their species, as do the substances volatilized by combustion. Without a basis yielding gelatine, there can be no true bone, but just as little can there be true bone without bone-earth, nor cartilage without cartilage-salts, nor blood without iron, nor saliva without *Potassium-chloride*.

"Man is generated of earth and air. The activity of plants called him to life. The corpse is decomposed into air and ashes, and through the vegetable world it then develops new forces in new forms."

These words caused me to found a biochemic therapy. The little work herewith submitted contains its development. In my biochemical theraphy only 11 remedies are used, these being such as are homogeneous with the inorganic substances contained in the blood and in the tissues of the human organism.

Owing to reasons which the reader will find on page 35 and those following, these remedies must be given in small doses.

Whenever small doses are mentioned, the reader usually at once thinks of Homœopathy; my therapy, however, is not homœopathic, for it is not founded on the law of similarity, but on the physiologico-chemical processes which take place in the human organism. By my method of cure the disturbances occurring in the motion of the molecules of the inorganic substances in the human body are directly equalized by means of homogeneous substances, while Homœopathy attains its curative ends in an *indirect* way by means of *heterogeneous* substances.

Some of my opponents have averred that those of my remedies, as *Silicic acid* and

Calcium phosphate, etc., which had already been used by physicians *before* biochemistry was established, are on that account not biochemical remedies. It would be just as correct or rather incorrect to assert that all remedies used *before* Hahnemann belong exclusively to allopathy. But the truth of the matter is this:

The principle according to which a remedy is selected stamps its impress upon it. A remedy selected according to the principle of similars is a homœopathic remedy, but a remedy which is homogeneous with the mineral substances of the organism, and the use of which is founded on physiological chemistry, is a biochemical remedy. A Homœopath using *Silicea* unconsciously acts biochemically. *Silicea* cannot produce any symptoms in a *healthy* person which could cause its use in diseases according to the principle of similars. Homœopaths use

Silicea on account of curative symptoms gained empirically. In the same way they act with respect to the cell-salts, which they used before the establishment of biochemistry.

DR. MED. SCHUESSLER.

Oldenburg, March, 1898.

THE CONSTITUENTS OF THE HUMAN ORGANISM.

Blood consists of water, sugar, fat, albumen, sodium chloride (common salt), potassium chloride, calcium fluoride, silicic acid (*Silicea*), iron*, lime, magnesia, soda and potash.

The latter are combined with phosphoric acid or with carbonic acid and sulphuric acid. Sodium salts predominate in the serum of the blood, potassium salts in the blood-corpuscles. Sugar, fat and the albumens are the so-called organic constituents of the blood; water and the above mentioned salts are the inorganic parts. Sugar and fat are composed of carbon, and hydrogen and oxygen; the albumens

---

*Manganese is not a constant constituent of the blood and is, therefore, an insignificant constituent so far as the formation of the cells is concerned.

consist of carbon, oxygen, hydrogen, nitrogen and sulphur.

Blood contains the material for all the various tissues, *i. e.*, the cells of the body. This material reaches the tissues through the walls of the capillaries, and thus makes good the waste in the cells caused by the transformation of its substances.

Sulphur, carbon and phosphorus are not found in a free state in the organism, but are always found as integral parts of organic combinations. Sulphur and carbon are found in albumen, carbon in the carbohydrates (*e. g.*, sugar and starch) and in the products resulting from the transformations of organic substances.

Phosphorus is contained in the lecithins and the nucleins. The sulphur contained in the albumen is oxidized into sulphuric acid by the inhaled oxygen, and this acid then combines with the bases of the carbonates into sulphates, while the carbonic acid is set free.

The albumen destined to build up new cells is split up, through the influence of oxygen, within the tissues. The products of such a division are the substances forming muscles, nerves, gelatine, mucus, keratin and elastin.

The substance which forms gelatine is intended for the connective tissue, for the bones, the cartilage and the ligaments; the substances forming mucus, muscles and nerves are destined for the mucus-cells, the muscle-cells, the nerve-cells, and the cells of the brain and the spinal marrow; the keratin is intended for the hair, the nails and the cells of the epidermis and the epithelium; the elastin for the elastic tissues. While this division takes place, mineral substances are set free. These serve to cover the deficiencies occurring in the cells owing to their functions or through pathogenic excitation ; they also serve, especially the phosphate of lime, to incite the formation of cells.

Those mineral substances, however, which are liberated in consequence of the retrogressive metamorphosis of the cells leave the organism by the ways appointed for excretion, thus forming products of disintegration.

During the retrogressive metamorphosis of the cells, their organic substances are finally transmuted into urea, carbonic acid and water. As these final products with the liberated salts leave the tissues, they make room for the organic substances which have not yet been thus transmuted, so that these also may pass to their final transmutation.

The products of retrogression are conveyed through the lymphatics, the connective tissues and the veins to the gall-bladder, the lungs, the kidneys, the bladder and the skin and removed from the organism together with the urine, perspiration, fæces, etc.

With respect to the significance of the connective tissue, we find the following in Moleschott:

"It is one of the noblest fruits of modern research, for the acquisition of which Virchow and von Recklinghausen have cleared the way, that the connective tissue has advanced from the indifferent part first assigned to it into an unlooked-for fruitful activity. That, which formerly seemed only intended to fill up or to form a protective covering now appears to us as the matrix through which the most secret currents pass from the blood to the tissues and back from these to the blood vessels, at the same time serving as one of the most important breeding places for young cells, which may then be raised from their undeveloped youthful form into the most special structures of the body."

When through means of food and drink, properly digested, the blood is compensated for the losses which it has suffered from supplying the nutritive material to the tissues, and when thus there is present in the tissues the nutritive material in the requisite quantity and in the right place, and when there is no disturbance in the motion of the molecules, then the building of *new* cells and the destruction of the *old* cells as well as the elimination of waste products proceeds normally, and the man is in a state of health.

When a pathogenic irritation touches the cell, its function is thereby at first increased, because it endeavors to repel this irritation. But when, in consequence of this activity, it loses a part of its mineral materials for carrying on its function, then it undergoes a pathogenic change. Virchow says: "The essence of disease is the cell changed pathogenetically."

Suppose the functional material lost in the contest with the pathogenic irritation to be, *e.g.*, *Potassium chloride*, then it has also lost a corresponding quantity of fibrin, for *Potassium chloride* and fibrin have a physiologico-chemical relationship. If the cell in its contest with the pathogenic irritation has lost *Calcium phosphate*, it has also lost a corresponding quantity of albumen, because *Calcium phosphate* has a similar relation to albumen as *Potassium chloride* has to fibrin. An exudation of fibrin, therefore, presupposes a deficiency of *Potassium chloride*, and an exudation of albumen presupposes a deficiency of *Calcium phosphate* in the cells immediately contiguous to the exudation referrred to. Losses in the other cell-minerals may be deduced from their several characteristics which may be perused below.

The cells which have undergone pathogenic changes, *i. e.*, the cells in which there is a deficiency in one of their mineral constituents, need a compensation by means of a homogeneous mineral substance. Such a compensation may be made spontaneously, *i. e.*, through the curative effort of nature, whereby the requisite substances enter the cells from their interstices. But if the spontaneous cure is delayed, therapeutic aid becomes necessary. For this purpose the required mineral substances are given in a molecular form. The molecules enter through the epithelium of the cavity of the mouth and throat into the blood and diffuse themselves in every direction. Those molecules which enter the seat of the disease enter there into a lively molecular motion, which communicates itself to the homogeneous substances around. These substances enter the cells which have undergone pathogenic changes, and thence a cure is affected. The cells, which have been restored to their integrity are then able to move again independently and thence to eliminate foreign substances or in general anything redundant, and therefore also any exudations that may be present.

The constitution of the cell depends on the constitution of the nourishing soil immediately surrounding it, just as the prosperous growth of a plant depends on the quality of the soil within the reach of the fibres of its roots. The agricultural chemist speaks of "*the law of the minimum*," according to which the nutritive substance of which there is a minimum in the soil must be supplied as the manure required for the plant. The agricultural chemist uses for this only three substances as manures, either nitrogen in combination (ammonia), or *Calcium phosphate* or potassa. The other nutritive substances required by the plant are contained in sufficient quantities in the soil.

"*The law of the minimum*" is also applicable to the biochemical substances. To give an example:

In the nourishing soil of the bones in a child suffering from rhachitis in consequence of disturbance in the motion of the molecules of *Phosphate of lime*, there has arisen a deficiency in this salt. The quantity of *Phosphate of lime* intended for the bones, which cannot reach its destination, would become redundant in the blood, but that it is excreted with the urine. For the kidneys have the function of providing for the right constitution of the blood, and, therefore, of excreting every foreign and every *redundant* constituent.*

After the disturbance in the molecular motion of the nutritive soil in question has been equalized by means of minimal doses of *Phosphate of lime*, the redundant *Phosphate of lime* may find its way into the normal current and the cure of the rhachitis may thus be effected.

The biochemical method supplies the curative efforts of nature with the natural material lacking in the parts affected, *i. e.*, the inorganic salts. Biochemistry endeavors to correct the physiological chemistry when it has deviated from its normal state. Biochemistry in a direct mode

*The liver together with the kidneys have the common function of caring for the constant constitution of the blood. But despite a normal constitution of the blood in general, nevertheless, in the immediate nutritive soil of a complex of cells, *i. e.*, in the nutritive fluid between the cells, there may arise a deficiency as to a certain salt, and a consequent disturbance in the molecular motion. This disturbance may prevent the entrance of the requisite salt from the blood into the cellular interstices.

reaches its end, which is: supplying a deficiency. The other curative methods which use means which are *heterogeneous* to the substances constituting the human organism reach this end in an indirect way.

Anyone who will consider without prejudice the end to be attained and the ways and means, will come to see that the biochemical remedies, when used after proper selection, are sufficient for the cure of all diseases curable by internal remedies.

Some physicians have asserted that the biochemical remedies ought to be proved on healthy persons, and their indications should be derived from the symptoms ascertained from such provings. But this is altogether erroneous. The indications of biochemical remedies must be derived from physiological and pathological chemistry, *i. e.*, through the results of their use in the various diseases.

Who can believe, that by giving large or small doses of the cell-salts to healthy persons, we could cause morbid symptoms having any similarity with puerperal fever, with typhoid fever, with articular rheumatism, with chills and fever, with hygroma patellæ, etc., etc.?

The biochemical remedies are used in minimal doses. The possibility of the action of small doses is manifest from the following:

Nature operates only by means of atoms and by means of groups of atoms or molecules. The growth of animals and of plants is effected by adding new atoms or groups of atoms to the molecular masses already collected. That infinitesimal, imponderable particles of substance still may operate in the organism, can not be contested when we consider that waves of light, which of a certainty are also imponderable, nevertheless cause molecular motions in the living, green parts of plants, by means of which carbonic acid is decomposed into carbon and oxygen, and that these same waves on photographic plates, as also in the delicate membrane of the retina, cause molecular motions, which cause the production of an image.

The use of small doses for the cure of diseases in the biochemical method is a chemico-physiological necessity. If we desire to convey into the blood, *e. g.*, some Glauber's salt, this is effected not by giving a *concentrated* solution of it. This would only act within the intestinal canal, causing a watery diarrhœa, and with these evacuations it would leave the organism. A *diluted* solution of Glauber's salt will enter the blood from the buccal and thoracic cavity and it will also thus enter into the other intercellular fluids, and, owing to the peculiarity of the salt, in that it attracts to itself water, it will cause the withdrawal of the redundant water in the tissues into the venous blood, and it will thus cause an increase in the secretion of urine.

Every biochemical remedy must be thus attenuated, so that the functions of the healthy cells may not be disturbed, and yet the functional disturbances present may be equalized.

In healthy men, animals and plants the salts are present in dilutions corresponding to about the 3d, 4th and 5th decimal medicinal dilutions. This may appear from the following analysis of the blood-cells in the human organism :

In 1000 grammes of blood-cells we find contained the following quantities of inorganic matter :

Iron	0 998
Potassium sulphate	0.132
Potassium phosphate	2.343
Potassium chloride	3.079
Sodium phosphate	0.633
Soda	0.344
Calcium phosphate	0.094
Magnesium phosphate	0.060

(See *Bunge's Lehrbuch der physiologischen und Pathologischen Chemie*, "Manual of physiological and pathological chemistry, p. 219.)

In 1000 grammes of the intercellular fluid (plasma) we find the following quantities of inorganic matter:

Potassium sulphate	0.281
Potassium chloride	0.359
Sodium chloride	5.545
Sodium phosphate	0.271
Soda	1.532
Calcium phosphate	0.298
Magnesium phosphate	0.218

(Vide Bunge Manual.)

Besides these, the intercellular fluid

contains Glauber's salt in minute quantities, with fluorine and *Silicea.*

With these analyses compare that of milk:

One litre (1000 grammes) of milk contains of inorganic matter the following quanties:

Potassa	0.78
Soda	0.23
Lime	0.33
Magnesia	0.06
Iron	0.004
Phosphoric acid	0.47
Chlorine	0.44

(Vide Bunge's Manual, p. 97.)

Milk also contains traces of fluorine and *Silicea.*

A litre of milk (1000 grammes = 15,443 grains) is the average quantity consumed daily by a suckling weighing about 6 kilogrammes. Now, if 6 centigrammes of *Magnesia* are sufficient to supply the daily call for *Magnesia* in a suckling, how small ought the dose to be to cure a neuralgia caused by an infinitesimal deficit in this salt in a minute part of the nervous tissue?

The amount of mineral substance contained in *one* cell is infinitesimal. By weighing, measuring and calculating, the physiologian, C. Schmidt, has computed that *one* blood-cell contains about the one-billionth part of a gramme of *Potassium chloride.* The one-billionth part of a gramme corresponds to about the 12th degree of decimal dilution.

Also allopathic remedies are effective in minute doses: Prof. Dr. Hugo Schulz, in Greifswalde says: Corrosive sublimate in a dilution of one part to 600,000, or even up to 800,000, causes a very violent fermentation, far exceeding the normal, in a solution of grape-sugar containing yeast. Particulars of this may be found in the

Berliner Klinische Wochenschrift, Nov. 4th, 1889.

In determining the dose of a biochemical remedy, the quantity of a morbid product cannot be considered as the determining factor. A very minute deficit of common salt may, *e. g.* cause in the cells of the epithelial layer of a serous sac a very copious serous exudation, and a compensation of molecules of common salt corresponding to this minute deficit may cause the reabsorption of this exudation.

A physician who wishes to use biochemical remedies can select his dose according to the quantitative relation here laid down.

In my practice I generally use the 6th decimal trituration.* In acute cases take every hour or every two hours a quantity

**Ferrum phosphoricum, Silicea and Calcium fluoride* I usually give in the 12th trituration.

of the trituration as large as a pea, in chronic cases take as much, three or four times a day, either dry or in a teaspoonful of water.

A milligramme of substance is calculated to contain an average of 16 trillions of molecules, the 6th decimal trituration should therefore contain about 16 billion molecules. This number is more than sufficient to equalize the disturbance in the molecular motions of the tissues.

The objection might be made that the molecules of the salts given as a medicine will unite themselves with the homogeneous salts contained in the blood, and the intended curative effect will thereby become illusory. But to this we should answer that the unition dreaded does not take place, because the carbonic acid in the blood serves as an isolating medium to the molecules of the salt.

The inorganic substances which serve

as a means in plants for their nutrition and for performing their functions are also taken up by them only in minimal quantities. Liebig says: "The strongest manuring with phosphates in a coarse powder is hardly to be compared in its effects with a far smaller quantity in an infinitesimal state of comminution, which causes a particle of the phosphate to be present in every part of the soil. A single root-fiber requires but an infinitesimal quantity of nutriment from the spot where it touches the soil, but it is necessary for its function and existence that this minimum should be present in that very spot." (See *Liebig's Chemische Briefe* (*Chemical Letters*), Vol. II, p. 295.) Minerals insoluble in water, if contained within the sphere of nourishment of the plant, must be dissolved by the acid juices contained in the root-fibres before they can enter the organism of the plant.

Mineral matter which enters man's stomach is exposed to the action of the hydrochloric acid contained in the gastric juice. If the mineral substance is, *e. g.*, a salt of iron, then a chloride of iron is formed in the stomach. If, therefore, it is desired to convey *Phosphate of iron* (*Ferrum phosphoricum*) to the cells that have suffered a pathogenic change, this must not enter the stomach. It must therefore, be given in a minimal dose; the remedy must be so far attenuated, that its liberated molecules may be able to enter into the blood through the epithelium of the mouth, the throat and the œsophagus and through the walls of the capillaries.

All substances indissoluble in water must be reduced to at least the 6th degree of the decimal scale of attenuation; substances which will dissolve in water can penetrate the above mentioned epithelial cells even in a lower attenuation.

In the 3d edition of the *Baeder Almanach* for 1886 we find on p. 121 the following remark:

"Judging from the results, and from the present analysis, the Rilchinger Water contains especially also those constituents with which, according to Dr. Schuessler's "Abridged Therapy," all curable diseases are cured by the biochemical method."

In the Rilchinger water some mineral substances are present in such minute quantities that, *e. g.*, *Phosphate of Magnesia* corresponds to the 8th attenuation, *Potassium chloride* to about the 5th and silicic acid to about the 6th decimal attenuation.

In the Balneologic letters of Prof. Beneke we read the following:

"We would lay especial stress on *one*

relation: This is the degree of concentration in which the solutions of salt are offered to the body. I am convinced that many of the most celebrated mineral springs afford their favorable results just by the fact that the effective ingredients are presented in such a very *attenuated* form, and the experience is very essential, that we frequently obtain the *most signal* effects through doses which, according to our usual ideas, are very *minute.* *

It is better, in prescribing a salt for a biochemical purpose, to make the dose too small than too large. If it is too small, the goal will be reached by repeating it; but if it is too large, the end to be gained is wholly lost.

The motto, "*Much will help much,*" rests on a traditional error, which can be-

---

* We cannot, however, recommend the use of mineral waters from the standpoint of biochemistry. Biochemical remedies are to be prescribed singly; mixtures are inadmissible.

come harmful by its effects; *e. g.*, large doses of iron, after spoiling the stomach, are evacuated, unassimilated, with the fæces, without touching the disease that ought to be healed through iron.

Those physicians who believe that large doses are required, but, at the same time, have little confidence in their medicines, when they themselves fall ill, do not take any medicine at all. In dosing other people but not themselves with their pills and mixtures, they remind us of the plantation-lord who said: "Down south we raise excellent field-peas; we cannot, indeed, eat them; but they are excellent for the negroes."

CHARACTERISTICS OF THE BIO-CHEMICAL REMEDIES.

IRON.

Iron and its salts possess the property of attracting oxygen. The iron contained in the blood-corpusles takes up the inhaled oxygen, thereby supplying with it all the tissues of the organism. The sulphur contained in the blood-corpuscles and in other cells, in the form of sulphate of potassa, assists in transferring oxygen to all the cells containing iron and the sulphate of potassa. When the molecules of iron contained in the muscle-cells have suffered a disturbance in their motion through some foreign irritation, then the cells affected grow flaccid. If this affection takes place in the annular fibres of the blood vessels, these are dilated; and as a consequence the blood contained in them is augmented. Such a state is called hyperæmia from irritation; such a hyperæmia forms the first stage of inflammations. But when the cells affected have been brought back to the normal state by the therapeutic effect of iron (*Phosphate of iron*) then the cells are enabled to cast off the causative agents of this hyperæmia, which are then received by the lymphatics in order that they may be eliminated from the organism.

When the muscular cells of the intestinal villi have lost molecules of iron, then these villi become unable to perform their functions: diarrhœa ensues.

When the muscular cells of the intestinal walls have lost molecules of iron, then the peristaltic motion of the intestinal canal is retarded, resulting in an inertia with respect to the evacuation of the fæces.

From the above, we deduce the following indications for iron:

When the muscular cells which have grown flaccid through loss of iron receive a compensation for their loss, the normal tensional relation is restored: the annular fibers of the blood vessels are shortened to their proper measure, the capacity of these vessels again becomes normal, and the hyperæmia disappears, and in consequence the inflammatory fever ceases.

Iron will cure:

1. The first stage of all inflammations.
2. Pains } caused by hyperæmia.
3. Hemorrhages } caused by hyperæmia.
4. Fresh wounds, contusions, sprains, etc., as it removes the hyperæmia.

The pains which correspond to iron are increased by motion, but relieved by cold.

In the muscle-cells, iron is found in the form of a phosphate; we should therefore in therapeutics use *Ferrum phosphoricum.**

PHOSPHATE OF MAGNESIA.

Phosphate of magnesia is contained in the blood-corpuscles, in the muscles, in the brain, in the spinal marrow, in the nerves, in the bones and the teeth.

When the motion of its molecules in the nerves is disturbed, there arise pains, also cramps and paralysis. The pains thence resulting are usually shooting like lightning flashes, or boring; often combined with a sensation of constriction or alternating therewith; they are at times roaming pains. They are amelior-ated by warmth and by pressure, aggra-vated by a light touch.

Phosphate of magnesia will cure head-

---

*As to the potency, I usually give the 12x trituration.

ache, face-ache, toothache and pains in the limbs if they are of the kind described above; so also cramps in the stomach, pains in the abdomen usually radiating from the umbilical region, relieved by hot drinks, by bending double, and by pressing on the abdomen with the hand, sometimes accompanied with watery diarrhœa.

It will also cure spasms of various kinds; spasms of the glottis, whooping cough, lock-jaw, cramps of the muscles of the calves, hiccough, tetanus, St. Vitus' dance, spasmodic retension of the urine, etc.

Further particulars concerning *Magnesia* may be found under "SCROFULOSIS AND TUBERCULOSIS."

CALCIUM PHOSPHATE (CALCAREA PHOSPHORICA).

Calcium phosphate is found in all cells; it is most abundant in the osseous cells (osseous corpuscles). It plays a most important part in the formation of new cells. It therefore serves as a remedy in anæmic states, and for the restoration of tissues after acute diseases. It is particularly applicable in cases where the formation of bones is delayed, as in rhachitis and craniotabes, so also when there is a defective ossification of a parietal bone, when the fontanels remain open too long, etc. It hastens the formation of callus in fractured bones, and also hastens dentition. In the latter case it competes with *Calcium fluoride.*

When the molecular motion of *Calcium phosphate* is disturbed in the epithelial cells of the serous sacs, there ensues a sero-albuminous effusion into these sacs. In this way arises the hygroma patellæ, the hydrops genu, etc. If these losses are compensated by minimal doses of *Calcium phosphate*, then the effusions are reabsorbed.

When the cells of the epidermis have lost *Calcium phosphate*, then albumen appears on the surface and dries there into a crust; this crust can be made to come off by doses of molecules of *Calcium phosphate.*

When the epithelium of a mucous membrane is diseased from the loss of *Calcium phosphate*, an albuminous secretion ensues which is cured by *Calcium phosphate.*

Calcium phosphate also cures spasms and pains caused by anæmia. Such pains are accompanied with formication, or a sensation of numbness or cold.

POTASSIUM PHOSPHATE (KALI PHOSPHORICUM).

Potassium phosphate is contained in the cells of the brain, the nerves, the muscles and the blood (the blood corpuscles), as also in the plasma (serum) of the blood and in the other intercellular fluids.

A disturbance in the motions of its molecules produces:

1. In the domain of the cells of thought: Despondency, anxiety, fearfulness, an inclination to weep, homesickness, suspiciousness, agoraphobia, weakness of the memory and similar ill humor.

2. In the vasomotory nerves: At first a small and frequent pulse, later on it is retarded.

3. In the sensory nerves: Pains with sensation of paralysis.

4. In the motory nerves: Weakness of the muscles and the nerves, even to paralysis.

5. In the trophic fibres of the *Nervus sympathicus:* Retardation of nutrition even to a total cessation thereof in a lim-

ited cellular area, and thence a softening and decay of the affected cells.

All changes in the state of health have the characteristic of depression.

Potassium phosphate cures states of depression of the mind and of the body, hypochondriac and hysterical ill humor, neurasthenia, nervous insomnia, spasms caused by so-called irritable weakness; also paralyses, septic states, septic hemorrhages, noma, scurvy, scurvy of the mouth, phagedenic chancre, carbuncles, typhoid fever, and typhous, adynamic states; progressive atrophy of the muscles; the round ulcer of the stomach, because this is caused by a disturbance of the function of the trophic fibres of the *sympathicus;* so also the alopecia areata (not to be confounded with herpes tonsurans). Also in the alopecia areata the cause is found in a disturbance of the functions of the trophic fibers of the sympathicus.

POTASSIUM CHLORIDE (KALIUM CHLORATUM) K Cl.

(Not to be confounded with Chlorate of Potash K Cl O_3.)

Potassium chloride is contained in almost all the cells, and is chemically related to fibrine. It will dissolve white or grayish-white secretions of the mucous membranes and plastic exudations. It is, therefore, the remedy for catarrhs when the secretion has the form described above; it is also the remedy for croupous and diphtheritic exudations. It answers also to the second stage of inflammation in serous membranes when the exudation is plastic.

When the cells of the epidermis lose molecules of *Potassium chloride* in consequence of a morbid irritation, then the fibrine comes to the surface as a white or whitish-gray mass. When dried, this forms a mealy covering. If the irritation has seized upon the tissue under the epidermis, then fibrine and serum are exuded, causing the affected spot on the epidermis to rise in blisters. Similar processes may take place in and below the epithelial cells.

NATRUM MURIATICUM. SODIUM CHLORIDE (COMMON SALT).

The water which is introduced into the digestive canal in drinking or with the food enters into the blood through the epithelial cells of the mucous membrane by means of the common salt contained in these cells and in the blood, for salt has the well-known property of attracting water. Water is intended to moisten all the tissues, *i. e.* cells. Every cell contains soda. The nascent chlorine which is split off from the *Natrum muriaticum* of the intercellular fluid combines with this soda. The *Natrum muriaticum* arising by this combination attracts water. By this means the cell is enlarged and divides up. Only in this way can cells divide so as to form additional cells.

If there is no common salt formed in the cells, then the water intended to moisten them remains in the intercellular fluids, and hydræmia results. Such patients have a watery, bloated face; they are tired and sleepy and inclined to weep. They are chilly, suffer from cold extremities and have a sensation of cold along the spine. At the same time they have a strong desire for common salt. (The cells deficient in salt cry for salt.) The common salt, of which they consume comparatively large quantities, does not heal their disease, because the cells can only receive the common salt in very attenuated solutions.

The redundant common salt present in

the intercellular fluid may in such cases cause such patients to have a salty taste in their mouth (an irritation of the nervus glossopharyngeus and the N. lingualis), and the pathological secretions of the mucous membranes as also of excoriations of the skin may be corrosive (salt-rheum).

The common salt acting in the healthy epithelial cells of the serous sacs regulates the passage of water from the arterial blood into these sacs. A functional disturbance of the molecules of common salt is followed by a transfusion of water into these sacs. If this disturbance is therapeutically removed by minimal doses of common salt, then these cells are enabled to reabsorb the water exuded.

A disturbance in the molecular motions of the molecules of common salt in the epithelium of the lachrymal or salivary glands is followed by lachrymation or salivation.

When an irritation affecting a dental branch of the trigeminus is transferred by means of the secretory fibers of the sympathicus to the epithelial cells of the salivary glands, so as to disturb the function of the molecules of common salt in these cells, then there is toothache with salivation.

The epithelial cells of the mucous membrane of the intestinal canal, by means of the common salt they contain, effect the transfer of the water drunk as a beverage into the blood of the vena-portæ. A disturbance of their function, through an irritation foreign to it, causes a reverse current: serum enters into the intestinal canal, causing a watery diarrhœa. If the irritation affects at the same time the mucous cells of the intestines, there arises a diarrhœa of water and mucus.

The mucin of the mucous cells appears on the surface as a glassy, transparent mucus. If the mucous cells contain too little common salt and too little mucin, then the natural secretion of mucus is depressed below the normal.

The carbonic acid contained in the blood by its voluminal effect liberates chlorine from the common salt contained in the epithelial cells of the peptonic glands. The soda thus set free combines with the carbonic acid and this combination passes into the blood, while the chlorine liberated combines with hydrogen and, dissolved in water, it enters the stomach as hydrochloric acid. Now when the epithelial cells of the peptonic glands are deficient in salt, and in consequence there is no hydrochloric acid formed, then the alkaline mucus secreted by the superficial epithelium of the mucous membrane of the stomach increases and we have catarrh of the stomach, eventually accompanied with vomiting of mucus.

In consequence of a considerable disturbance of the function of the common salt, serum from the blood may transude into the stomach, then there arises vomiting of water (water-brash).

When a number of cells *below* the epidermis contain no common salt, they cannot receive the water destined for them; then they will raise up the epidermis in the form of vesicles; the contents of these vesicles are *clear as water*.

Similar vesicles may rise from a similar cause on the conjunctiva.

There may be simultaneously, though in places distant from one another, diminished or increased secretions in consequence of the disturbance in the function of the molecules of common salt; *e. g.*, there may be catarrh of the stomach with vomiting of water or of mucus, and at the same time constipation from a diminished secretion of mucus in the colon.

SODIUM PHOSPHATE (NATRUM PHOSPHORICUM).

Sodium phosphate is contained in the blood-corpuscles, in the cells of the muscles, of the nerves and of the brain, as well as in the intercellular fluids. Through the presence of *Sodium phosphate*, lactic acid is decomposed into carbonic acid and water. *Sodium phosphate* is able to bind to itself carbonic acid, receiving into itself two parts of carbonic acid for every volume of phosphoric acid. When it has thus bound the carbonic acid, it conveys it to the lungs. The oxygen flowing into the lungs liberates the carbonic acid which is only loosely attached to the *Sodium phosphate;* the carbonic acid is then exhaled and exchanged for oxygen, which is absorbed by the iron of the blood-corpuscles.

Sodium phosphate is the remedy for those diseases which are caused by an excess of lactic acid. It, therefore, answers to the diseases of infants, who, having been fed to excess with milk and sugar, suffer from redundant acids. The symptoms in such cases are: Sour eructations, vomiting of sour, cheesy masses; yellowish-green, so-called hacked diarrhœas; colic, spasms with acidity.

Uric acid is dissolved in the blood by two factors; the warmth of the blood and *Sodium phosphate*. If uric acid is deposited from its solution in the joints or near them, owing to a deficiency of *Sodium phosphate*, or when it combines with the base of *Carbonate of soda* into urate of soda which is insoluble, then there arises podagra or acute arthritic rheumatism. During an acute attack of podagra the secretion of uric acid in the urine is diminished by just so much as is retained of it in the diseased parts.

Sodium phosphate also serves to saponify the fatty acids; it, therefore cures those dyspeptic ailments which arise from eating fat food, or which are aggravated thereby.

Additional facts concerning *Sodium phosphate* will be found under "SCROFULOSIS AND TUBERCULOSIS."

CALCIUM FLUORIDE (CALCAREA FLUORATA).

Calcium fluoride is found in the surface of the bones, in the enamel of the teeth, in the elastic fibers and in the cells of the epidermis. A disturbance in the motion of its molecules with a consequent loss thereof is followed:

1. By a hard, lumpy exudation on the surface of a bone.

2. By a relaxation of elastic fibers; thence an enlargement of the vessels, hemorrhoidal knots; relaxation and change of position of the uterus, relaxation of the abdominal coverings, sagging down of the abdomen; the after-pains are deficient, or there may be hemorrhages from the uterus.

3. The Keratin* or horny substance exudes from the cells of the epidermis. The exudate dries up at once and becomes a crust, firmly adhering to the skin; it thus appears *e. g.*, on the palms. When the hand thus affected is used, chaps and tears in the crusts are formed.

Besides these diseases, *Calcium fluoride* will cure:

a. Cephalæmatom; since it causes the absorption of the osseous wall.

b. Hardened exudations, *e. g.*, in the mammary glands, the testes, etc.

---

*Keratin is contained in the epidermis, the hair and the nails.

Two explanations may be offered as to the absorption of hardened exudations:

a. Through the pressure of the hardened exudation, the elastic fibers near it have lost the ability of performing their function. By supplying molecules of *Calcium fluoride*, the affected fibers are restored to their integrity and are thence enabled to throw off the exudation, which is then reabsorbed by the lymphatics.

b. Through the voluminal action of the carbonic acid in the blood, a part of the fluorine in the *Calcium fluoride* is split off. This detached fluorine combines then with nascent hydrogen into hydrofluoric acid, which gradually dissolves one molecule of the exudate after the other, and these are then received by the lymphatics.

The sulphuric acid formed by the oxidation of the albuminous corpuscles may at times play the part of the carbonic acid in liberating the fluorine.*

SILICIC ACID, (SILICEA).

Silicic acid is a constituent of the cells of the connective tissue, of the epidermis, the hair and the nails.

If a suppurative centre is formed either in the connective tissue or in a portion of the skin, *Silicea* may be used.

After the functional ability of the cells of the connective tissue, which had been impaired by the pressure of the pus has been restored to its integrity through a supply of molecules of *Silicea*, these cells are thereby enabled to throw off inimical substances (the pus). In consequence, the pus is either absorbed by the lymphatics or it is cast out. In the latter case there is a so-called spontaneous breaking open of the suppurative center.

Silicea may also cause the absorption

*As to the potency, I give the 12x trituration.

through the lymphatics of an effusion of blood in any tissue. If the reabsorption of a sero-albuminous exudation in a serous sac cannot be effected through *Calcarea phosphorica*, then *Silicea* may be used; for the delay in the absorption may also be caused by a deficiency in *Silicea* in the subserous conjunctive tissue.

Silicea will also cure chronic arthritic-rheumatic affections, as it forms a soluble combination (*Sodium silicate*) with the soda of the ureate of soda; this combination is then absorbed and removed through the lymphatics. For the same reason it may also be used in renal gravel.

Silicea can also restore the perspiration of the feet when this has been suppressed, and is thus an indirect remedy in diseases arising in consequence of such suppression (*e. g.*, amblyopia, cataract, paralysis, etc).

When a number of cells in the conjunctive tissue are gradually deprived of *Silicea*, they become atrophied. Such a disease is by no means rare in the external meatus auditorius with old people. The meatus in such a case is dry and enlarged.*

THE SULPHATES.

The sulphuric acid formed during the oxidation of the albuminous corpuscles would destroy the tissues, if this acid did not in its nascent state combine with the bases of carbonates of the alkalies (potassa and soda) liberating their carbonic acid.

SODIUM SULPHATE (NATRUM SULPHURICUM).

The action of the *Sodium sulphate* is contrary to that of the *Sodium chloride*. Both, indeed, have the faculty of attracting water, but the end is a contrary one; the *Sodium chloride* attracts the water des-

*As to the potency, I generally give the 12x trituration.

tined to be put to use in the organism, but the *Sodium sulphate* attracts the water formed during the retrogressive metamorphosis of the cells, and secures its elimination from the organism.

The *Sodium chloride* causes the splitting up of the cells necessary for their multiplication; the *Sodium sulphate* withdraws water from the superannuated leucocytes and thus causes their destruction. The latter salt is, therefore, a remedy for leukæmia. *Sodium sulphate* is a stimulant of the epithelial cells and of the nerves, as will appear in what follows.

In consequence of the activity excited by *Sodium sulphate* in the epithelial cells in the urinary canals, superfluous water with the products of the tissue changes, dissolved or suspended therein, flows into the kidneys, in order to leave the organism in the form of urine through the ureters and the bladder.

While the *Sodium sulphate* stimulates the epithelial cells of the biliary ducts, the pancreatic ducts and of the intestines, it causes the secretion of the excretions of these organs.

Sodium sulphate is also intended to stimulate the functions of the nerves of the biliary apparatus, of the pancreas and of the intestines.

If the *sensory* nerves of the bladder are *not* stimulated by *Sodium sulphate*, the impulse to void urine does not come to man's consciousness; thence there follows involuntary micturition (wetting the bed).

If the *motory* nerves of the detrusor are not stimulated, there results retention of urine.

In consequence of an *irregular* action of the *Sodium sulphate* on the epithelial cells and the nerves of the biliary apparatus, there arises either a diminution or an increase of the secretion and excretion of the bile.

If the motory nerves of the colon are not sufficiently influenced through *Sodium sulphate*, there arise constipation and flatulent colic.

If in consequence of a disturbance in the motion of the molecules of *Sodium sulphate* the elimination of the superfluous water from the intercellular spaces takes place too slowly, there arises hydræmia, and functional disturbances in the apparatus for the secretion of bile cause the following diseases:

Chills and fever, bilious fever, influenza, diabetes, bilious vomiting, bilious diarrhœa, œdema, œdematous erysipelas; on the skin, vesicles containing yellowish-water, moist herpes, herpes circinnatus, sycotic excrescences, catarrhs with yellowish-green or green secretions, etc.

The state of health of persons suffering from hydræmia is always worse in humid weather, near the water, and in damp, moist under-ground dwellings; it is improved by contrary conditions.

POTASSIUM SULPHATE (KALI SULPHURICUM).

Potassium sulphate, which in reciprocal action with iron effects the transfer of the inhaled oyxgen to all the cells, is contained in all the cells containing iron.

Where there is a deficiency as to *Potassium sulphate*, according to the locality and extent of the deficiency, the following symptoms may arise:

A sensation of heaviness and weariness, vertigo, chilliness, palpitation of the heart, anxiety, sadness, toothache, headache and pains in the limbs. These ailments increase while the person is confined to a room, also in the warmth and toward evening, and they are relieved in the open, cool air.

There ensues a desquamation of cells of the epidermis and the epithelium, which have been loosened from their connection because they were not sufficiently provided with oxygen. The scaling off of these epithelial cells is followed by catarrhs with a secretion of yellow mucus.

Therapeutically, *Potassium sulphate* answers to the process of desquamation which takes place after scarlatina, measles, erysipelas of the face, etc.

It also cures laryngeal catarrh, and catarrhs of the bronchia, of the conjunctiva, of the mucous membrane of the nostrils, etc., where the secretion has the above mentioned characteristics; also a catarrh of the stomach, when the tongue has a yellowish mucous coating; also a catarrh of the middle ear and renal catarrh.

Potassium sulphate effects the access of oxygen, and this hastens the formation of new cells of the epidermis and of the epithelium, whereby the cells that have been loosened from their connection are thrown off.

Also in inorganic nature, sulphates and iron serve for the transfer of oxygen. When in the surface layer of the earth a sulphate and any oxide of iron come into contact with organic substances undergoing decomposition, they surrender their oxygen and form sulphuret of iron. This may be again decomposed through the access of new oxygen, so that sulphuric acid and some oxide of iron will be formed, which under suitable conditions will again transfer their oxygen.

CALCIUM SULPHATE.

In Moleschott's *Physiologie der Nahrungsmittel* (Physiology of nutriments)

Calcium sulphate is enumerated as a nutriment. This work was published as long ago as the year 1859. Since that time many views have been corrected.

In Bunge's Manual of Physiological and Pathological Chemistry, which appeared in the year 1887, *Calcium sulphate* is found *only* in analyses of the bile and only in two of these, while it is not found in two others (pp. 189, 190).

On page 23 of his Manual, Bunge says of sulphur: "It enters into the bodies of animals chiefly in the form of albumen, and, after the decomposition and oxidation of albumen, it issues again for the most part in the highest stage of oxidation, as sulphuric acid. In this form, in combination with alkalies, it leaves the animal body to begin its cycle anew."

Sulphuric acid is thus combined in the body, not with earths, with calcium and magnesium, but with alkalies, with potassa and soda.

Calcium sulphate has, indeed, been successfully used in many diseases, (in suppurative processes, and in affections of the skin and of the mucous membrane); but as it may be seen from the above quotations that it does not enter into the constant constitution of the organism, it must disappear from the biochemical system.

Instead of it *Sodium phosphate* and *Silicea* are to be considered.

The inorganic substances found in the blood and in the tissues suffice for the cure of all diseases that are at all curable.

Chronic diseased states, produced by the abuse of medicines, such as quinine, mercury, etc., can be cured by minimal doses of cell-salts.

The symptoms determine the choice of the remedies.

While the above mentioned diseases caused by medicines can be cured with cell-salts, acute cases of poisoning with arsenic, phosphorus, etc., must, of course, be treated according to the well-known principles that have reference thereto.

Several physicians have asserted that the organic combinations found in the human organism must also be received into the biochemical therapy. But this idea is founded on error, as I shall endeavor to show.

Biochemical therapy is, as we have already indicated above, an analogue to agricultural chemistry. If a plant possesses the inorganic substances naturally belonging to it, it is able to form of itself all the organic combinations which its organism needs. We do not manure the plants with grains of chlorophyllum in order that we may cause their leaves to become green, for we know that the iron contained in the plants will provide the green for the leaves. We do not manure with lecithin, nuclein, etc., to provide the plants with these combinations of phosphorus; if necessary, we manure with phosphate of lime. The plant takes from the calcium phosphate the phosphoric acid, and combines this with the substances present within them which are necessary for the formation of lecithin, nuclein, etc.

If anyone should assert that agricultural or horticultural chemists are mistaken in thinking that three kinds of manure are sufficient, and should say, that all the organic substances found in plants must be considered in providing a manure, *e. g.*, chlorophyllum, gum, resin, oil, starch, grape-sugar, malic acid, etc., one would merely smile at "Daniel come to judgment."

If the human organism contains organic

nutriments, such as albumen, fat and carbohydrates, together with the proper inorganic cell-salts, in sufficient quantities in the right place, then, through the influence of oxygen and in consequence of decompositions and syntheses, all the necessary organic combinations will arise, and the individual in question will be in a state of health.

Syntheses, which were formerly thought to be a peculiar privilege of the vegetable kingdom, take place as well in the human and animal organisms.

Among those who think that organic substances should also be received into my biochemical system is also Dr. Ring of Ward's Island, New York. He finds fault with me, because I have not received the *original combinations* of organic substances into my system. He says among other things: "Organic substances, like keratin, tyrosine, creatine, creatinine, etc., are normal constituents of those substances in which and upon which cancerous swellings are formed, and we are therefore justified in supposing that, if rightly prepared and rightly chosen, they should exert a specific action on the tissues related to them."

This is in part true, but for the greater part erroneous. It is true that keratin is a normal constituent of some tissues; but it is not correct to say that creatine and creatinine are constituents of the tissues; they are merely contained in them as the products of the retrogressive metamorphosis of the cells. All organic combinations which, like creatine, creatinine, urea, uric acid, etc., are excreted in normal urine are to be considered as the final stages of the oxidation of organic nutriments. As to their uselessness to the human organism, they may be compared with the resin which is excreted by some plants as a product useless for them.

The idea that we might cure a diseased tissue with a related sound tissue is curious. The cartilaginous tissue is related to the mucous tissue. *Natrum muriaticum* is the functional agent in each of them. Now would any one cure a coryza, a disease of the mucous tissue that may be cured by means of *Natrum muriaticum*, with a preparation of cartilage?

A number of years ago, Dr. Constantine Hering had the idea of proving the horny tissue as a remedy. He and his friends prepared *Castor equorum*, the horny excrescence on the legs of horses, and proved it both on horses and on men. In the list of symptoms we find the statement: "An old decrepit horse became 20 years younger." Despite this symptom which promised so much, and which stamps *Castor equorum* as something analogous with the wonderful mill which is to make old women young again, the remedy has, nevertheless, sunk into oblivion.

Dr. Ring and his associates are having the substances above-mentioned prepared, and will try their effects on healthy persons. Their undertaking will give rise to manufactories of symptoms. We shall probably hear of very amusing symptoms.

If the chemico-physiological views of these gentlemen were a little clearer, they would see that their undertaking is a useless diversion.

If they should, *e.g.*, use lecithin, they can at most discover the effects of a phosphate; if they should make provings of keratin which is very rich in sulphur, they can at most find out the effects of a sulphate. Why roam afar, when biochemistry already contains five phosphates and two sulphates?

If an inorganic salt is abundantly excreted in the urine, then, owing to a disturbance in the molecular motion, there will be found a deficiency as to the same salt in the immediate nourishing soil of some cellular domain, and a homogeneous salt is indicated as the remedy (*vide* Rhachitis, p. 32); a minimum in the nourishing soil is always a cell-salt, never an organic substance, therefore, organic substances are excluded from our remedies. Whoever may doubt this, can try whether any disease can be cured through the molecules of gelatin, mucus, tyrosine, elastin, fat, sugar, etc. The result will ever be a negative one.

For the construction and preservation of the human organism, the following substances are required: Oxygen, fat, albumen, a gelatinous substance, mucin, keratin, elastin, hæmoglobin, lecithin, nuclein, cholesterin, water and inorganic salts.

Albumen forms the chief constituent of the plasma of the blood and of the lymph; it is contained in the muscular fibers, the cylinders of the axis of the nervous fibers and in the protoplasma of all cells. The organic frame of the bones, the cartilage, the fascia and the connective tissue consist of a gelatinous substance. Mucin is contained in the epithelial cells of the mucous membranes. Keratin is the organic basis of the epidermis, the hair and the nails; elastin is the basis of the elastic fibers.

The gelatinous substance, mucin, keratin and elastin, are products of the splitting up of albumen under the influence of oxygen.

The hæmoglobin of the blood-cells is the combination of a corpuscle of albumen with hæmatin, a corpuscle containing iron.

Lecithin and nuclein arise from albumen, fat and a phosphate, through a change of position of the molecules.

Whatever other organic and inorganic constituents may be found in the tissues are merely the products of the retrogressive metamorphosis of the cells and of the decomposition of the albumen; they are substances which must be eliminated through the activity of the cells.

Among these products of the retrogressive metamorphosis of the cells are also, as before said, creatine and creatinine; and among the products of the decomposition of albumen we find tyrosine, leucin, etc.

The albumens and the fats are means of supply and sources of power; oxygen, carbohydrates and gelatin (not to be confounded with the substances supplying gelatin) are also sources of power; the inorganic salts are means of supply and regulators of the functions.

Equalization of functional disturbances

is synonymous with the restoration of health. This end is sought for in the biochemical method only through inorganic salts.

The hope of Dr. Ring and his associates, that they may effect cures by means of keratin, creatine, etc., is founded on an illusion, which disappears when viewed in the light of physiology.

SPECIAL GUIDE FOR USING THE BIO-CHEMICAL REMEDIES.

FEVERS.

Fever is intended to effect the removal of the exciting agents of the disease, as also of its products. During fever the changes in the substances of the tissues are increased. By means of the fragments (scoriæ) resulting from the retrogressive transformation of the cells, both the exciting agents and the products of the disease are removed from the tissues and eliminated *through* the excretory channels. In such a way nature may effect a cure. But such a cure does not always ensue; therefore therapeutic aids are expedient. But whenever fever is depressed by means of antipyrin, antifebrin, quinine, etc., the changes in the substances and thereby also the cure are delayed. Nevertheless, this is done by many physicians; but such action is opposed to nature. The fact that despite of such an unnatural treatment many persons escape with their lives, simply shows that it takes a good deal to kill even a sick person *secundum artem.* Sometimes, of course, the effects are mischievous. So I read a short time ago in a paper from Southern Germany that a patient who had only had a *slight* attack of pneumonia had died, after his fever had been repressed by too large a dose of quinine. This case, which reminds us of the son of the sexton of Tweedledum, who could not digest opium, proves that the descendants of Dr. Eisenbart are not all dead as yet.

With respect to the *biochemical* treatment of fever, *Ferrum phosphoricum* corresponds to inflammatory fever, since it cures the exciting hyperæmia, which causes inflammatory fever. (See the Characteristics of the Effects of Iron, on page 48 .)

The fever which accompanies typhus, the puerperal fever and acute rheumatism of the joints diminish in proportion as these diseases are cured under the influence of *Kali phosphoricum*, *Natrum phosphoricum*, etc.

EXUDATIONS AND TRANSUDATIONS.

Exudation of fibrine: *Kali chloratum.*
" " albumen: *Calcarea phosphorica.*
" " clear water: *Natrum muriaticum.*
" " yellowish water: *Natrum sulphuricum.*
Exudation of mucus: *Natrum muriaticum.*

When the exudation becomes smeary and fetid : *Kali phosphoricum.*

If an exudation of mucus becomes yellowish (yellow mucus), then *Kali sulphuricum* will answer.

A phlegmonous inflammation of the skin or of the subcutaneous connective tissue requires *Natrum phosphoricum.* If a suppurative center is developed, *Silicea* is to be used, which sometimes effects the reabsorption of the pus, but in most cases causes the breaking of the abscess outwardly and thus effects a cure.

If the pus becomes fetid, *Kali phosphoricum* should be used; if indurations remain, *Calcium fluoride* is to be used.

INFLAMMATION OF THE SEROUS MEMBRANES.

Meningitis Pleuritis Pericarditis Endocarditis Peritonitis	*Ferrum phosphoricum* corresponds to the first stage. For further indications see EXUDATIONS.

PNEUMONIA AND PLEURITIS.

In the stage of hyperæmia: *Ferrum phosphoricum.* For further indications see EXUDATIONS.

ARTICULAR RHEUMATISM, PODAGRA, GOUT.

Natrum phosphoricum dissolves the *uric acid* accumulated in the affected parts, and thus makes it innocuous. It is then eliminated, together with the uric acid it has taken up, through the channels in the organism through which the transmutations of the substances are effected.

Deposits of *urates* requires *Silicea.* See the characteristics of *Silicea*, p. 69.

With respect to muscular rheumatism, see what is stated under the heading: "'Pains in the Neck, Back and Limbs."

DISEASES OF THE KIDNEYS.

The remedies corresponding to *inflammation of the kidneys*, are *Ferrum phosphoricum*, *Kali chloratum and Natrum phosphoricum.*

The remedies corresponding to *Albuminuria* are *Kali sulphuricum*, *Calcarea phos-*

phorica, *Kali phosphoricum* and *Natrum muriaticum.* The accompanying symptoms and the constitutional state of the patient must decide in the choice of the remedies.

Albuminuria after scarlatina, requires *Kali sulphuricum.* The epithelial cells of the uriniferous tubes, while in a *healthy* state, resist the pressure of the albumen of the blood, it is only the *diseased* cells which allow albumen to enter the uriniferous tubes. The epithelium mentioned may be diseased, owing to a lack of a sufficient supply of oxygen, or owing to their premature decay, or owing to the delay in the division and new formation of cells.

Silicea will prevent the formation of renal gravel.

PUERPERAL FEVER.

The specific remedy in this disease is *Kali phosphoricum.*

TYPHUS AND TYPHOID.

The specific remedy is *Kali phosphoricum.* In cases of deep stupor *Natrum muriaticum* is indicated as an accompanying remedy.

TYPHOID. ADYNAMIC SYMPTOMS.

When in an acute disease, accompanied with fever (diphtheritis, scarlatina, small-pox, etc.), sopor, dryness of the tongue, watery vomiting, etc., set in, *Natrum muriaticum* is useful. When the teeth show a brown coating, the evacuations have a cadaverous fetor, attended with septic hemorrhages, *Kali phosphoricum* will answer.

DIPHTHERIA.

The form which appears most frequently, the so-called catarrhal form, with slight swelling and a greyish-white exudation, answers to *Kali chloratum.* With considerable swelling and an abundant white exudation, which frequently also covers the uvula, *Calcarea phosphorica* is the suitable remedy. When the exudation on the swollen tonsils is yellow, *Natrum phosphoricum* is indicated.

If gangrene appears, *Kali phosphoricum* should be used. This remedy will also cure the paralytic states which frequently appear after diphtheria has run its course: a nasal voice, strabismus, etc.

The simultaneous use of lime-water, ice, carbolic acid, etc., is altogether objectionable, as also wrapping in wet cloths to produce perspiration. These means exhaust the strength of the patient. Children sometimes die from this weakness, as frequent experience has shown.

CROUP.

In pseudo-croup, *Kali chloratum* is indicated; in genuine croup, *Calcarea phosphorica.*

Owing to the specific relation existing between *Calcarea phosphorica* and albumen, molecules of this phosphate combine with the albuminous molecules of the lower surface of the croupous exudation adhering to the mucous membrane. In consequence of this process, the exudation separates from the mucous membrane. This separation of the exudation from the mucous membrane may be hastened by alternate doses of *Calcarea phosphorica* and of *Kali sulphuricum.* The latter transmits oxygen from the blood (see p. 75), and the oxygen favors the formation of new epithelial cells from the albumen separated from the croupous exudation. The molecular motions taking place during this process hasten the separation of the exudation.

The alternate use of *Calcarea phosphorica* and *Kali sulphuricum* is also in place in diphtheria with a *white* exudation.

DYSENTERY.

Ferrum phosphoricum and *Kalium chloratum* are in most cases sufficient. If delirium and distension of the abdomen set in, and the evacuations have a cadaverous smell, *Kali phosphoricum* is suitable. This remedy also answers when a copious quantity of pure blood is discharged, without any signs of putridity.

Spasmodic abdominal pains, relieved by pressure and by doubling up, require *Magnesia phosphorica.*

SCARLATINA.

In light cases, *Ferrum phosphoricum* and *Kalium chloratum* are sufficient. The remedy suitable to severe cases will be found by considering what is said under the head of DIPHTHERIA and TYPHOID AND ADYNAMIC SYMPTOMS.

Kali sulphuricum corresponds to the dropsy appearing after scarlatina.

SMALLPOX.

Kalium chloratum should be used in the beginning. If the pustules show pus, *Natrum phosphoricum* will be suitable. If symptoms of adynamia and decomposition of the blood arise, *Kali phosphoricum* should be given. In confluent pustules, *Natrum muriaticum* is required.

MEASLES.

The accompanying symptoms will indicate the remedy: *Ferrum phosphoricum*, *Kalium chloratum*, *Kali sulphuricum* and *Natrum muriaticum* are chiefly to be considered.

INFLUENZA.

The remedy in influenza is *Natrum sulphuricum* (see the characteristic of this salt, p. 71.).

The cases of influenza which I treated with *Natrum sulphuricum* showed no after-effects. The diseases left in cases where other physicians had treated influenza with other remedies were of such a nature that they were covered by the sphere of *Natrum sulphuricum*, therefore they could be cured with this remedy.

PAINS OF THE HEAD AND FACE.

Stitches, or pressure or beating, increased by shaking the head, by stooping and, in general, by every motion: *Ferrum phosphoricum.*

Pains, accompanied by heat and redness of the face: *Ferrum phosphoricum.*

Pains, with vomiting of bile: *Natrum sulphuricum.*

Pains, with vomiting of transparent mucus or water: *Natrum muriaticum.*

Pains, with vomiting of food: *Ferrum phosphoricum.*

Pain, with retching up of white mucus: *Kalium chloratum.*

Quick, shooting, lancinating pains, intermittent, and varying in their location: *Magnesia phosphorica.*

Pains in pale, sensitive, irritable persons: *Kali phosphoricum.*

Paroxysms of pain, followed by great debility: *Kali phosphoricum.*

Pains which are aggravated in a warm room and in the evening, but are relieved in the open, cool air: *Kali sulphuricum.*

Pains, accompanied by the simultaneous appearance of small nodules, the size of a pea, on the scalp: *Silicea.*

Pains, attended with a coating of clear mucus on the tongue and sluggish evacuations: *Natrum muriaticum.*

Pains, attended with a copious flow of acrid tears: *Natrum muriaticum.*

Disguised intermittent fever, appearing as neuralgia of the head or face: *Natrum sulphuricum* and eventually *Natrum muriaticum.*

The headaches of children are, as a

rule, quickly cured by *Ferrum phosphoricum.*

Pains, with formication and a sensation of coldness or numbness: *Calcarea phosphorica.*

SCALP.

The external application of *Natrum muriaticum* is useful in scab-head and in the falling out of the hair.

Alopecia areata: *Kali phosphoricum.*

Herpes tonsurans: *Natrum sulphuricum.*

CONCUSSION OF THE BRAIN.

Kali phosphoricum is the answering remedy. If disturbances of vision remain, *Magnesia phosphorica* is indicated.

Hydrocephaloid: *Calcarea phosphorica.*

Chronic hydrocephalus: *Calcarea phosphorica.*

Cephalæmatom: *Calcarea fluorata.*

Craniotabes: *Calcarea phosphorica.*

When the fontanelles remain open too long: *Calcarea phosphorica.*

If in any of these diseases there is diarrhœa with a cadaverous stench, *Kali phosphoricum* must be given as an intermediate remedy.

Apoplexy: *Silicea.*

DELIRIUM TREMENS.

Most cases of this ailment are rapidly cured by means of *Natrum muriaticum.* If this should fail, *Kali phosphoricum* should be given.

VERTIGO.

If vertigo is caused by a rush of blood, *Ferrum phosphoricum* should be given; if it is nervous, it will be cured by *Kali phosphoricum.* If there are any gastric troubles attending it, the coating of the tongue must be considered.

EARS.

Pains caused by hyperæmia, noises in

the ear, and difficulty in hearing, require *Ferrum phosphoricum.*

In nervous affections, *Magnesia phosphorica*, *Calcarea phosphorica* or *Kali phosphoricum* should be chosen, bearing a proper regard to the individualities.

Inflammatory swelling, closing the meatus auditorius externus: *Silicea.*

Discharge of thin, yellow fluid: *Kali sulphuricum.*

Discharge of thick pus: *Silicea*, *Natrum phosphoricum.*

Hardness of hearing, due to a swelling and to catarrh in the Eustachian tube and of the tympanic cavity: *Kalium chloratum*, *Natrum muriaticum.*

If there is reason to think that hardness of hearing is caused by indurated exudations in the interior ear, *Silicea* and *Calcarea fluorata* should be given.

Mumps: *Kalium chloratum*, and if

there is copious salivation, *Natrum muriaticum.*

TOOTHACHE.

Pains, attended with salivation or lachrymation: *Natrum muriaticum.*

Pains, with a swelling of the gums and cheek: *Kalium chloratum;* if this is insufficient: *Silicea;* if the swelling is hard like bone: *Calcium fluorata.*

Pain, which quickly changes its location, is intermittent, and is alleviated by warmth: *Magnesia phosphorica.*

Pain which is alleviated by pressure and worse when lightly touched: *Magnesia phosphorica.*

Pain which grows worse in a warm room and in the evening, but is alleviated in the open, cool air: *Kali sulphuricum.*

Hot cheeks, with increase of pain by warm drinks, alleviated by cold drinks: *Ferrum phosphoricum.*

If the gums bleed or have a bright reddish border: *Kali phosphoricum.*

If the painful tooth is loose, and its surface painful to the slightest touch: *Calcarea fluorata.*

AILMENTS DURING TEETHING WITH CHILDREN.

Calcarea phosphorica and more especially *Calcarea fluorata*, assist the coming through of the teeth.

When there is fever: *Ferrum phosphoricum.*

Spasms with fever: *Ferrum phosphoricum.*

Spasms without fever: *Magnesia phosphorica* and *Calcarea phosphorica.*

Inflammation of the eyes: *Ferrum phosphoricum* and *Calcarea phosphorica.*

Slavering: *Natrum muriaticum.*

Spasm of the larynx: *Magnesia phosphorica.*

Spasmodic cough: *Magnesia phosphorica.*

Spasm of the bladder: *Magnesia phosphorica.*

Diarrhœa, *vide* DIARRHŒA.

EYES.

Blepharitis ciliaris: *Kalium chloratum*, *Natrum phosphoricum.*

Styes, nodules, induration of the lids: *Silicea*, *Calcarea fluorata.*

Hyperæmia of the conjunctiva without any secretion: *Ferrum phosphoricum.*

When the secretion is white, grayish-white: *Kalium chloratum.*

When the secretion is watery mucus: *Natrum muriaticum.*

When the secretion is yellow mucus: *Kali sulphuricum.*

When the secretion is thick yellow, like pus: *Natrum phosphoricum*, eventually *Silicea.*

When the secretion is yellowish green : *Natrum sulphuricum.*

When the secretion is like cream: *Natrum phosphoricum.*

Inflammation of the eyes of the newborn: Chief remedy, *Natrum phosphoricum*; other biochemical remedies according to the secretion (to be given internally and also for squirting into the eyes).

Inflammation of the eyes in scrofulous persons: Chief remedies, *Natrum phosphoricum* and *Magnesia phosphorica.*

Trachoma: *Kalium chloratum.*

Inflammation of the cornea: *Kalium chloratum*, if the exudation is whitish-grey; *Calcarea phosphorica*, if it is white; *Natrum phosphoricum*, if it is yellow.

Vesicles on the cornea: *Natrum muriaticum.*

Flat ulcer on the cornea: *Kalium chloratum.*

Deep ulcer: *Silicea.*

Spots on the cornea: The spot is to be syringed several times a day with an attenuation of *Natrum muriaticum.* The molecules of *Natrum muriaticum*, adhering to the spot affected, produce, through their power of absorbing moisture, a gradual thorough moistening of the spot, and thence it will melt away.

Hypopyon: *Silicea.*

Inflammation of the iris: *Kalium chloratum, Natrum muriaticum.*

Inflammation of the retina: *Ferrum phosphoricum.*

Retinal exudation: *Kalium chloratum.*

Photophobia after over-exertion, without any other symptoms: *Kali phosphoricum.*

Fiery sparks before the eyes: *Natrum phosphoricum, Magnesia phosphorica.*

Spasmodic strabismus: *Magnesia phosphorica;* when caused by worms: *Natrum phosphoricum.*

Strabismus after diphtheria: *Kali phosphoricum.*

Nervous asthenopia: *Kali phosphoricum.*

Hydræmic asthenopia: *Natrum muriaticum.*

Violent boring pains in the eye, as a purely nervous affection: *Magnesia phosphorica;* as a rheumatic affection: *Natrum phosphoricum;* as an arthritic affection: *Silicea.*

Pains in the eyes with lachrymation, appearing daily at set times: *Natrum muriaticum.*

CAVITY OF THE MOUTH.

Catarrhal inflammation of the **mucous membrane** covering the soft palate the tonsils and the pharynx:

When redness and violent pain are present: *Ferrum phosphoricum.*

When there is a white exudation: *Kalium chloratum.*

When the exudation is golden yellow: *Natrum phosphoricum.*

When there is a transparent frothy mucus: *Natrum muriaticum.*

Angina tonsillaris: *Natrum phosphoricum*; to chronic swelling of the tonsils corresponds: *Magnesia phosphorica.*

Inflammation of the uvula: *Natrum muriaticum.*

Inflammation of the tongue: If the tongue is greatly swollen and dark red: *Ferrum phosphoricum.* Should suppuration set in: *Silicea.* For induration: *Calcarea fluorata.*

Cancrum oris and **scurvy**: *Kali phosphoricum.*

Gums: If the gums are pale, *Calcarea phosphorica* is most suitable. If the gums have *a bright red border*, *Kali phosphoricum* is indicated. The latter also answers when the gums bleed.

Coating of the tongue: For a white

coating, not mucous, *Kalium chloratum* is suitable. If the coating is mucous and on the edges of the tongue there are minute bubbles of mucous saliva: *Natrum muriaticum.*

If the tongue is clean and moist: *Natrum muriaticum.*

If the tongue has a dirty, brownish-green coating, attended with a bitter taste: *Natrum sulphuricum.*

If the tongue is, at it were, spread over with liquid mustard, attended with an offensive odor from the mouth: *Kali phosphoricum.*

Coating, golden yellow and moist: *Natrum phosphoricum.*

When the tongue has a yellow mucous coating: *Kali sulphuricum.*

The influence of the coating of the tongue in determining the choice of the remedy does not extend to the affections of all the tissues; but it is to be regarded in those cases which I have pointed out in this treatise. If any one suffering from chronic catarrh of the stomach has some other (acute) disease added thereto, the coating of the tongue will not always indicate the remedy for the acute disorder.

But when a disease—especially a chronic one—exhibits only uncertain symptoms, then the coating of the tongue will in most cases lead to the choice of the right remedy.

Aphthæ and **Thrush**: *Kalium chloratum*, when it is white or whitish-gray; but when yellow: *Natrum phosphoricum.* When there is a bright-red border: *Kali phosphoricum.*

Noma: *Kali phosphoricum.*

VOMITING.

Vomiting of food: *Ferrum phosphoricum.*

Vomiting of food together with a sour fluid: *Ferrum phosphoricum.*

Vomiting of bile only: *Natrum sulphuricum.*

Vomiting of transparent mucus, drawn out in long threads: *Natrum muriaticum.*

Vomiting of a watery fluid: *Natrum muriaticum.*

Vomiting of blood: *Ferrum phosphoricum*, *Kali phosphoricum* and *Natrum phosphoricum.*

Retching up of white mucus: *Kalium chloratum.*

Vomiting of a sour fluid or of cheesy masses: *Natrum phosphoricum.*

Vomiting during dentition: *Calcarea phosphorica*, *Calcarea fluorata.*

Sea-sickness: *Natrum phosphoricum.*

JAUNDICE.

The first remedy to be given in every case of jaundice is *Natrum sulphuricum.* This remedy will in most cases effect a cure. As a second resort we have

Kalium chloratrum, *Kali sulphuricum* and *Natrum muriaticum*, which should be selected according to the concomitant symptoms.

PAINS IN THE STOMACH AND ABDOMEN.

Acute inflammation of the stomach with violent pain of the distended gastric region, vomiting and fever: *Ferrum phosphoricum*.

If in a case where treatment has been delayed there are symptoms of exhaustion, dryness of the tongue, etc., *Kali phosphoricum* should be given.

Acute and chronic gastralgias, aggravated by eating and by pressure on the gastric region, and especially if food is vomited, require *Ferrum phosphoricum*.

Cramp-like gastrodynia, with clean tongue: *Magnesia phosphorica*.

Sensation of spasmodic constriction: *Magnesia phosphorica*.

Stomach-pains with gathering of water in the mouth: *Natrum muriaticum.*

Pains in the stomach, with vomiting of mucus, attended with indolent stool: *Natrum muriaticum.*

If *Natrum muriaticum* does not prove sufficient in these pains, there will usually be found a coating of the tongue, which calls for *Kalium chloratum* or *Kali sulphuricum.*

Pressure and **feeling of fulness**, while the tongue is coated with yellow mucus: *Kali sulphuricum.*

Pinching in the stomach with eructation of small quantities of air, affording no relief: *Magnesia phosphorica.*

Pains, caused by accumulation of flatus in the colon: *Natrum sulphuricum.*

Colic in the umbilical region, compelling the person to bend double: *Magnesia phosphorica.*

Flatulent colic of little children, with

drawing up of the limbs, with or without diarrhœa: *Magnesia phosphorica.* If there is an excess of acid, *Natrum phosphoricum* should be given.

In gastric pains accompanied with vomiting, the character of the matter vomited will indicate the remedy.

Gastric affections where acidity (heartburn) predominates: *Natrum phosphoricum;* also after fat food, *Natrum phosphoricum*, as it saponifies the fatty acids.

Ulceration of the stomach. The round ulcer of the stomach, which is caused by a disturbance in the function of the trophic fibers of the sympathicus, requires *Kali phosphoricum.*

Flatulent colic with constipation, in adults: *Natrum sulphuricum.*

Painters' colic: *Natrum sulphuricum* (2d dilut.).

Gall-stone colic (where a stone has entered the ductus choledochus and lodged there): *Magnesia phosphorica.*

Natrum phosphoricum may prevent the new formation of gall-stones.

Enlargement of the stomach: *Kali phosphoricum.*

DIARRHŒA.

Evacuations watery, mucous: *Natrum muriaticum.*

Evacuations of carrion-like fetor: *Kali phosphoricum.*

Evacuations, watery-bilious: *Natrum sulphuricum.*

Evacuations, bloody, bloody-mucous: *Kali chloratum.*

Evacuations, purulent, bloody-purulent: *Natrum phosphoricum*, eventually *Silicea.*

Evacuations undigested: *Ferrum phosphoricum.*

Diarrhœa caused by redundant acid: *Natrum phosphoricum.*

Watery diarrhœa with colic before every evacuation: *Magnesia phosphorica.*

Cholerine and cholera: *Natrum sulphuricum.*

WORMS.

Natrum phosphoricum is of use in the case of the oxyuris vermicularis, by destroying the excess of lactic acid which conditions the existence of these worms; for the ascaris lumbricoides, *Natrum muriaticum.*

HÆMORRHOIDS.

The remedy for hæmorrhoids is *Calcarea fluorata.* When the varices are inflamed *Ferrum phosphoricum* should be given. In violent pains, without inflammation, *Magnesia phosphorica* is suitable. In the so-called mucous hæmorrhoids, *Natrum muriaticum* is indicated.

DIABETES MELLITUS.

The remedy for this disease is *Natrum*

sulphuricum. A very prominent concomitant symptom outside of the sphere of *Natrum sulphuricum* may require a remedy corresponding to that symptom.

CORYZA.

Dry coryza: *Kalium chloratum;* with scrofulous persons: *Natrum phosphoricum.*

Fluent coryza: the secretion watery, of clear mucus: *Natrum muriaticum.*

Fluent coryza: the secretion a yellow mucus: *Kali sulphuricum.*

The secretion thick, purulent: *Natrum phosphoricum*, eventually *Silicea.*

In ozæna, *Natrum phosphoricum* and *Magnesia phosphorica* are useful.

When a *green* mucus is secreted, *Natrum sulphuricum* is indicated.

HOARSENESS.

In simple hoarseness arising from a cold, *Kalium chloratum* is suitable. It is seldom that *Kali sulphuricum* is required afterward. When the hoarseness is a consequence of over-exertion of the vocal organs (with actors, singers, etc.), *Ferrum phosphoricum* and eventually *Kali phosphoricum* will be useful.

COUGH.

An acute, short, spasmodic and very painful cough requires *Ferrum phosphoricum*, followed by *Kalium chloratum*. To genuine whooping cough corresponds *Magnesia phosphorica*. With respect to cough accompanied with an expectoration of mucus, see DISEASES OF THE MUCOUS MEMBRANES.

ASTHMA.

Kali phosphoricum and *Magnesia phosphorica* correspond to *nervous* asthma; the latter remedy in cases attended with flatulence.

Respiratory ailments connected with

catarrhal symptoms, *i. e.*, which are caused thereby, indicate the remedies required by the mucus expectorated. (*Vide* DISEASES OF THE MUCOUS MEMBRANES.)

WHOOPING-COUGH.

The inflammatory catarrhal stage requires *Ferrum phosphoricum*, the nervous stage *Magnesia phosphorica*. In the vomiting of food, *Ferrum phosphoricum* is useful. According to the quality of the mucus, *Kalium chloratum*, *Natrum muriaticum* or *Kali sulphuricum* are to be selected.

A *special concomitant symptom* may call for the use of an inter-current remedy (*e. g.*, *Kali phosphoricum*, *Calcarea phosphorica*).

ACUTE ŒDEMA OF THE LUNGS.

Dyspnœa, blueness of the face, convulsive cough, with the expectoration of a frothy-serous mass, require *Kali phosphoricum* and *Natrum muriaticum*.

DISEASES OF THE MUCOUS MEMBRANES.

In selecting the remedy, the consistence and color of the secretion are decisive:

If fibrinous: *Kalium chloratum.*

If albuminous: *Calcarea phosphorica.*

If golden-yellow: *Natrum phosphoricum.*

If yellowish, mucous: *Kali sulphuricum.*

If green: *Natrum sulphuricum.*

If clear, transparent: *Natrum muriaticum.*

If purulent: *Natrum phosphoricum, Silicea.*

If very fetid: *Kali phosphoricum.*

If excoriating: *Natrum muriaticum* and *Kali phosphoricum.*

The remedies for coughs with expectoration, leucorrhœa, coryza, catarrh of the frontal sinuses, etc., should be selected on the basis of the above distinctions.

POLYPUS.

When the gelatinous substance which forms the organic foundation of the connective tissues loses *Phosphate of lime*, there may result thence a loosening and a spongy excrescence of the tissue in question. When a part of the sub-mucous connective tissue is diseased through loss of *Phosphate of lime*, a polypus is formed. This may be cured by *Calcium phosphate*.

CATARRH OF THE BLADDER.

The chief remedy to be considered is *Natrum phosphoricum*.

DISEASES OF THE MUCOUS MEMBRANES.

Silicea generally corresponds to chronic catarrh of the bladder.

Hypertrophy of the prostate gland: *Magnesia phosphorica*.

RETENTION OF URINE (ALSO WETTING THE BED).

From the characteristics of the effects

of *Natrum sulphuricum* (p. 71) it appears that this remedy may cure as well the retention of urine as also involuntary micturition (wetting the bed). Should, however, the one or the other ailment be caused by a general or a local neurasthenia, then *Kali phosphoricum* should be used.

In strangury caused by a spasm of the sphincter, *Magnesia phosphorica* is useful.

In children suffering from worms, *Natrum phosphoricum* should be given to prevent wetting the bed.

The retention of urine in little children, attended with heat, is cured by *Ferrum phosphoricum.*

DISEASES OF THE SKIN.

The remedies recommended in diseases of the mucous membranes also correspond to affections of the skin: eczema, herpes, etc.

Vesicles with sero-fibrinous contents: *Kalium chloratum.*

Vesicles with albuminous contents: *Calcarea phosphorica.*

Vesicles with watery-clear contents: *Natrum muriaticum.*

Vesicles with honey-yellow contents: *Natrum phosphoricum.*

Vesicles with yellowish-watery contents: *Natrum sulphuricum.*

Vesicles with puriform contents: *Natrum phosphoricum* or *Silicea.*

Vesicles with bloody, ichorous contents: *Kali phosphoricum.*

Pustules with pus on an infiltrated base: *Silicea.*

The scabs, scales or crusts appearing after the bursting of the vesicles require the following remedies:

Mealy scurf: *Kalium chloratum.*

Yellowish-white crusts: *Calcarea phosphorica.*

White scales: *Natrum muriaticum.*

Honey-yellow crusts: *Natrum phosphoricum.*

Yellowish scales: *Natrum sulphuricum.*

Yellow, purulent crusts: *Silicea.*

Fetid, greasy crusts or scales: *Kali phosphoricum.*

Profuse scaling off of the epidermis on a viscid base: *Kali sulphuricum.*

Hard crusts on the palms, with or without chaps: *Calcarea fluorata.*

Swelling of the sebaceous glands: *Natrum phosphoricum.*

Inflammation and suppuration of these glands: *Silicea.*

The humid eruptions call for the *Natrum salts*, varied according to the varying colors of the secretions mentioned above.

For eruptions arising after vaccination, *Kalium chloratum* or *Natrum phosphoricum* should be used.

For excoriation of infants: *Natrum phosphoricum* and *Natrum muriaticum.* If attended with a diarrhœa of cadaverous odor, use *Kali phosphoricum.*

Urticaria or nettle-rash: *Kali phosphoricum.*

Pruritus: *Magnesia phosphorica.*

Rhagades or chaps: *Calcarea fluorata.*

Disorders in the nails of the fingers; when they break easily, tear, become yellow, have spots or grow thick: *Silicea.*

Erysipelas.—The œdematous, soft inflammation of the skin requires *Natrum sulphuricum;* to the infiltrated inflammation *Natrum phosphoricum* corresponds.

For herpes zoster, *Natrum muriaticum* should be used.

In erysipelatous inflammations, symptoms of intense fever and inflammation may indicate *Ferrum phosphoricum. Kali sulphuricum* will further the desquamation.

Pemphigus.—The pemphigus vulgaris (bullæ and vesicles with watery contents and fully distended surface) requires *Natrum sulphuricum* if the fluid is yellowish; but if the fluid is clear, like water: *Natrum muriaticum.* To pemphigus malignus (blisters and vesicles with watery-bloody contents and flaccid and wrinkled surface) corresponds *Kali phosphoricum.*

Burns and scalds: When a blister has been formed, give *Natrum muriaticum.* If there is an open surface covered with a white or grayish-white exudation, give *Kalium chloratum.* If suppuration has already ensued, *Silicea* is suitable. These remedies should be applied both internally and externally.

Chilblains, *fresh* and *suppurating*: *Natrum sulphuricum.*

Panaritium: *Silicea.*

Furuncle: *Silicea.*

Carbuncle: *Calcarea fluorata*, later *Kali phosphoricum*.

Proud flesh: *Kalium chloratum*, eventually *Silicea*.

Consequences of the stings of insects: *Natrum muriaticum* (externally).

Warts on the hands: *Kalium chloratum*. A quantity of the trituration, the size of a pea, should be dissolved in a tablespoonful of water; with this solution moisten the warts and the surrounding skin several times a day.

Also *Natrum sulphuricum* may be used. It withdraws the water from the base of the warts and thereby causes them to become flaccid and to fall off.

MASTITIS.

Natrum phosphoricum should first be used; if given in time, it may cause a reabsorption. If suppuration has set in, *Silicea* is to be used. Induration: *Calcarea fluorata*.

LYMPHATIC GLANDS.

See the paragraph on **Scrofulosis** and **Tuberculosis.** So also what is said in various passages on "SUPPURATION" and "INDURATION."

GOITRE.

Magnesia phosphorica.

CHANCRE AND GONORRHŒA.

The soft chancre requires *Kalium chloratum*, but the phagedenic chancre, *Kali phosphoricum;* the hard chancre, *Calcarea fluorata.* These remedies should be used internally and externally.

For chronic syphilis, *Kalium chloratum*, *Kali sulphuricum*, *Natrum muriaticum*, *Natrum sulphuricum*, *Silicea* and *Calcarea fluorata* should be used, according to the symptoms.

Gonorrhœa: The chief remedy is *Natrum phosphoricum.*

In bleeding of the urethra *Kali phosphoricum* is useful.

For gleet *Natrum muriaticum* and *Calcarea phosphorica* should be used.

If the secretion is greenish or green, give *Natrum sulphuricum.*

Condylomata require *Kalium chloratum* and *Natrum sulphuricum.*

Orchitis calls for *Ferrum phosphoricum*, then *Kalium chloratum*, and eventually *Calcarea phosphorica.*

Induration of the testicles: *Calcarea fluorata.*

Œdema of the scrotum: *Natrum muriaticum*, and *Natrum sulphuricum.*

Œdema of the prepuce: *Natrum muriaticum* and *Natrum sulphuricum.*

Balanitis: *Kali sulphuricum;* if fetid, use *Kali phosphoricum* (externally and internally).

Hydrocele: *Natrum muriaticum*, *Calcarea phosphorica*, eventually *Silicea.*

MECHANICAL INJURIES.

Contusions, incised and other fresh wounds, sprains, etc., require at once *Ferrum phosphoricum.* If, after the use of this remedy, a swelling remains, give *Kalium chloratum.* If, in neglected cases, suppuration ensues, *Silicea* is suitable. In sanious discharge or gangrene: *Kali phosphoricum;* proud flesh: *Kalium chloratum.*

Fractures of bones require, besides the mechanical measures, at first *Ferrum phosphoricum* for the lesion of the soft parts; later *Calcarea phosphorica* to promote the formation of callus.

Tenalgia crepitans (crepitating or crackling painful tendons), an ailment arising above the wrist on the dorsal side of the forearm with joiners and other artisans, as a result of over-exertion in using a chisel or other tool with a semi-rotatory motion, has been quickly relieved by me in two cases with *Ferrum phosphoricum.*

A third case which under allopathic treatment had become chronic, I quickly cured with *Kalium chloratum*, after *Ferrum phosphoricum* had refused to act.

Ganglium tendinosum: *Calcarea fluorata.*

ULCERS OF THE LEGS.

In such cases the remedies recommended for diseases of the skin and of the mucous membranes are to be considered.

First of all should be mentioned *Natrum muriaticum* and *Natrum sulphuricum.*

For varicose ulcers use *Calcarea fluorata.*

DISEASES OF BONES.

Periostitis with a tendency to suppuration requires *Silicea.*

Hard, knobby, jagged elevations on the surface of the bone require *Calcarea fluorata.*

This remedy will be also found more suitable than *Silicea* in cephalæmatoma, a bloody tumor with an osseous wall on the parietal bone in newborn children.

Rickets require *Calcarea phosphorica.* If attended with atrophy and a fetid diarrhœa, this condition must first be removed by means of *Kali phosphoricum.* Excessive acidity must be eliminated by *Natrum phosphoricum.*

Dr. Kassowitz, in Vienna, Prof. Hagenbach, in Bern, and others prescribe in rickets phosphorus in minimal doses.

The recipe in question is as follows:

℞.		
	Phosphori, ...	0.01
	Solve in ol. amygd. dulc., ...	10.0
	Pulv. gumm. arab.,	
	Syr. simpl., ... āā	5.00
	Aqu. distill., ...	80.00

This mixture represents the *fourth* decimal attenuation of phosphorus; but as it is given in teaspoonful doses, the daily quantity given is about equal to the usual *third* decimal dilution. The molecules of phosphorus in such a case combine within the organism with molecules of oxygen into phosphoric acid. This combines with the molecules of *Carbonate of lime* with the elimination of carbonic acid into *Phosphate of lime.* Such a treatment of rickets agrees both as to quantity and quality with the treatment given in this book, when *Calcarea phosphorica* is given in the 3d decimal trituration.

Since a part of the molecules of phosphorus or of the phosphoric acid on the way to its destination has the opportunity of combining with the molecules of soda in the blood, the cells in question will perhaps receive only a part of the dose of phosphorus destined for them. The possibility that the soda may appropriate *all* the molecules of the phosphorus furnished, explains the occasional failures in this treatment. But if *Calcarea phosphorica* is prescribed, a surer result will be attained, as this will not combine with the above-mentioned salts.

Inflammation of the hip-joint in scrofulous persons: *Natrum phosphoricum* and *Silicea.*

HÆMORRHAGES.

Blood, red, easily coagulating into a gelatinous mass: *Ferrum phosphoricum.*

Blood, black, thick, viscid: *Kalium chloratum.*

Blood, bright-red or blackish-red, at the same time thin and watery, not coagulating: *Kali phosphoricum* and *Natrum muriaticum.*

To epistaxis in children, *as a rule*, corresponds *Ferrum phosphoricum.*

For the predisposition to epistaxis give *Kali phosphoricum.*

Uterine hæmorrhages: especially *Ferrum phosphoricum*, *Calcarea fluorata* and *Kali phosphoricum.*

Bleeding from hæmorrhoids: *Ferrum phosphoricum*, *Kalium chloratum* and *Calcarea fluorata*.

MENSTRUATION.

In disturbances of the menstrual function, the accompanying symptoms must decide the choice of the remedy.

LABOR.

Labor-pains, weak: *Kali phosphoricum*; spasmodic labor pains: *Magnesia phosphorica*.

Deficient labor pains: *Calcarea fluorata* when the relaxation of the elastic fibers of the uterus is the cause, but *Kali phosphoricum* when there is deficient innervation.

MENSTRUAL COLIC.

Usually: *Magnesia phosphorica*. Pale, sensitive, irritable persons, inclined to weep, require *Kali phosphoricum*.

If accompanied with accelerated pulse

and increased redness of the face: *Ferrum phosphoricum.*

Vaginism: *Ferrum phosphoricum, Magnesia phosphorica.*

SECRETION OF MILK.

Natrum sulphuricum diminishes the secretion of milk, *Calcarea phosphorica* increases it.

Natrum muriaticum should be used when the milk is bluish and watery.

PAINS IN THE BACK OF THE NECK, THE BACK AND THE LIMBS.

Pains which are only felt during motion, or are aggravated by motion, require *Ferrum phosphoricum* (as a second remedy *Kalium chloratum* is suitable).

Pains, laming, ameliorated by moderate exercise, but made worse by a fatiguing effort (as by long-continued walking) and most felt when beginning to move, as when rising from a seat: *Kali phosphoricum.*

Pains with sensation of numbness or of cold, or with formication, worse at night and while at rest: *Calcarea phosphorica.*

Pains quick, shooting, boring, intermitting, changing their place: *Magnesia phosphorica.*

Pains, worse in the warm room and toward evening; better in the open, cool air: *Kali sulphuricum.*

In pains which the patients cannot exactly describe, some other attendant symptom which may decide the selection, such as an eruption of vesicles, the color of the coating of the tongue, etc., should be discovered.

Crick in the back: *Ferrum phosphoricum, Natrum phosphoricum.*

Pains in the hip: nervous pains require *Kali phosphoricum* and *Magnesia phosphorica* (to be selected according to the variety of the pain); inflammatory pains: *Ferrum phosphoricum;* rheumatic-arthritic pains: *Natrum phosphoricum*; if chronic: *Silicea.*

Hygroma patellæ and Hydrops genu require *Calcarea phosphorica*; eventually *Silicea* is to be used.

SPASMS AND OTHER NERVOUS AFFECTIONS.

In palpitation of the heart, *Ferrum phosphoricum*, *Kalium chloratum*, *Natrum muriaticum*, *Kali phosphoricum*, *Kali sulphuricum*, etc., are called for, according to the symptoms accompanying each case.

The chief remedies in epilepsy are: *Kali chloratum*, *Natrum muriaticum*, *Natrum phosphoricum*, *Kali phosphoricum* and *Magnesia phosphorica.* They are to be selected according to the characteristics before given.

Nocturnal paroxysms require *Silicea.*

Calcarea phosphorica corresponds to the spasms of anæmic and rachitic persons.

Spasms of the glottis, tetanus, trismus, cramp in the calves of the legs, writers' cramp, St. Vitus' dance, etc., require *Magnesia phosphorica*, *Calcarea phosphorica* and *Kali phosphoricum*.

Kali phosphoricum corresponds to cramps arising from an over-exertion of the parts affected.

Agoraphobia: *Kali phosphoricum*.

INTERMITTENT FEVER.

Natrum sulphuricum and *Natrum muriaticum* are the remedies for intermittent fever.

Natrum sulphuricum stands first; but *Natrum muriaticum* is suitable when an eruption of vesicles on the lips or some other symptom indicating common salt is present.

Natrum sulphuricum cures by killing redundant leucocytes, by withdrawing the water from them, and by eliminating from the organism the redundant water resulting from the retrogessive transformation of the cells.

Natrum muriaticum owes its curative powers to the fact that it increases the number of red blood-corpuscles and effects a proper distribution of the water necessary to the tissues.

Patients with intermittent fever should not eat any fat viands.

SCROFULOSIS AND TUBERCULOSIS.

It is well known that sugar of milk, which is a constituent of milk, is changed into lactic acid by a so-called ferment, and also that lactic acid causes a coagulation of the albumen contained in the milk. It is also known that *Natrum phosphoricum* decomposes lactic acid into carbonic acid and water. These facts serve to explain the formation of swellings of the lymphatic glands, when lactic acid is present there, as also the curability of such swellings of the lymphatic glands by means of *Natrum phosphoricum*.

When there is a redundancy of lactic acid in the organism and a portion of this lactic acid gets into the lymphatic glands, then a coagulation of the albumens in the lymph within these glands takes place and we have swellings of the lymphatic glands. These swellings, so long as they have not become indurated, may be removed by *Natrum phosphoricum*, because this salt decomposes the lactic acid, as before said, into carbonic acid and water. When the lactic acid is decomposed, the albumen not yet indurated becomes fluid again and can then enter again into the lymphatic current.

Since the lymph also contains fat, the coagulated albumen may also be saponified. If there is a caseous degeneration in the glands or in other places, *Magnesia phosphorica* is to be used.

But so long as there is not as yet any caseous degeneration, we should use *Natrum phosphoricum*, as may be seen from what is said above; but caseous degeneration requires *Magnesia phosphorica.* This is the chemico-physiological functional remedy which secures the independent activity of all the cells. Owing to their independent motion, *sound* cells are able to reject substances which encumber them. When the cells near these caseous masses are too weak to reject them, they are deficient in *Magnesia phosphorica.* By the therapeutical supply of minimal quantities of this salt these cells are restored to their integrity and thus enabled to gradually reject these tuberculous masses. The detritus of the rejected masses is then removed from the organism by the usual excretive channels.*

Magnesia phosphorica has proved its

---

* *Magnesia phosphorica* is perhaps also a cure for cancer.

efficacy in tuberculosis not too far advanced and in lupus.

Besides the use of *Magnesia phosphorica*, the use of other biochemical remedies is required to cure the catarrhal symptoms and the hæmorrhages from the lungs, etc.

What is the relation of the bacilli to tuberculosis? When there are tubercles, bacilli can come in and use them for their nourishment. What mites are in old cheese, bacilli are to the tubercles.

CHLOROSIS AND OTHER ANÆMIC STATES.

The blood-corpuscles contain, as has been shown in the analysis on page 38, iron, *Potassium sulphate*, *Potassium chloride*, *Potassium phosphate*, *Phosphate of lime*, *Phosphate of magnesia*, *Sodium phosphate* and soda. The multiplication of blood-corpuscles is effected through their division while in the current of the blood; it is effected in the following manner:

From the *Sodium chloride* contained in the plasma of the blood, a portion of chlorine is split off through the carbonic acid acting in volume; the part split off, combines with the soda contained in the corpuscles into *Sodium chloride.* This attracts serum and receives it into itself; thereby the corpuscles are enlarged and in consequence they subdivide. The small cells issuing from this division take up blood-albumen to effect their growth, and this is organized by means of *Phosphate of lime.*

In the blood-albumen the iron necessary for the formation of blood-corpuscles is present in sufficient quantity; in the normal (red) blood-cell the proportion in weight of the iron to the cells is as 1 to 1000. (*Vide*, p. 38.)

When in a blood-cell there is a minimum of soda, no sufficient quantity of *Sodium chloride* can be formed, as may appear from the above statement; the contents of water in the blood-corpuscle cannot then be increased in the degree necessary for its partition.

If there is a minimum of *Phosphate of lime* in the intercellular fluid, then the albumen necessary for the growth of the young cells cannot be organized in sufficient quantity. In cases where *Sodium chloride* cannot be formed in the cells, this salt must be furnished to the patients in minimal doses. The *Sodium chloride* of the intercellular spaces represents a solution of common salt which is too much concentrated for the cells; it must therefore be given in a higher dilution.

If the common salt of the blood could enter into the diseased and into the healthy blood-corpuscles, their partition would be effected prematurely; for partition would follow on partition even to eventual annihilation, for the small cells resulting from these partitions would have no time for their growth and for entering on their functions.

Sodium chloride and *Phosphate of lime* are the chief remedies in chlorosis. If we cannot in any stated case determine exactly which of the two remedies is indicated, the two remedies may be given in alternation.

Anæmic states which have been caused by depressing emotions need for their cure *Kali phosphoricum*, because this salt is then contained in a minimum quantity in the blood-corpuscles and in the plasma of the patient. The general state of health of the patient or at least some of his symptoms will be imaged forth in the characteristics of *Kali phosphoricum* (*Vide*, p. 54.)

The remedy for leukæmia is *Natrum sulphuricum*, which causes the disintegration of the superannuated leucocytes by withdrawing water from them. Therefore it will also cure chronic suppurations in leukæmic patients.

SOME THOUGHTS AS TO THE ALLOPATHIC TREATMENT OF CHLOROSIS.

Some allopaths now use lime in their treatment of chlorosis. By doing this they have unconsciously entered into the paths of biochemistry. As chlorotic patients frequently have a desire for chalk—*Carbonate of lime*—we may say: the diseased cells cry out for lime! This voice of nature ought to have been hearkened to before this.

The great number of allopathic physicians, as is well known, open their campaign against chlorosis with iron. The use of this remedy for this disease is as old as the history of medicine. The fact that all known preparations of iron have been used in the treatment of this disease, and that other and better preparations are still being sought after, shows that all these various curative efforts have not as yet satisfied anybody. The rejection of known preparations of iron and the search after new ones in order to cure chlorosis have been nothing but moving in a vicious circle. Iron and its artificially concocted combinations do not enter from the intestines into the blood.

Every salt of iron introduced into the stomach is decomposed there. "They are all transformed," as Bunge says in his Manual of Physiological and Pathological Chemistry, p. 91, "into combinations with chlorine. When these touch the walls of the stomach, which are always alkaline from *Sodium carbonate*, the chloride is transformed into an oxide, which remains in solution owing to the presence of organic substances. The chlorate of iron is transformed into carbonate of oxide of iron, which is also soluble in the carbonic acid and the organic substances present. Its not being absorbed is not, therefore, a consequence of its insolubility. Finally the combinations of iron being acted upon by the various combinations of sulphur and the reducing agencies—of the nascent hydrogen and other products of partition which are readily oxidized—they are changed into sulphuret of iron and excreted with the fæces. The combinations of iron with organic acids cannot act otherwise. Among the organic acids we must also number the albumens. The iron albuminates are also at once decomposed by the hydrochloric acid in the gastric juice, forming chlorides and chlorates of iron. Our food must, therefore, contain quite different combinations of iron, combinations which are not destroyed in the intestinal canal, which are absorbable and furnish the material for the hæmoglobin."

From this it is plainly manifest that iron and the artificial combinations of iron cannot by their direct action cure chlorosis.

Those allopathic physicians who are now using muriatic acid to cure chlorosis obtain thereby better results than those who are unwilling to give up iron. Muriatic acid favorably affects the digestion in the stomach, but iron spoils the stomach when it is given in allopathic doses. When the peptonic glands of chlorotic patients do not furnish sufficient muriatic acid, the function of the stomach is depressed.

Bunge says, on page 95 of his manual: "The main significance of the gastric juice consists probably in the antiseptic action of the free muriatic acid. When the quantity of this muriatic acid is insufficient, then fungi and bacteria enter into the intestinal canal, especially also those producing the fermentation of butyric acid. But in this fermentation hydrogen is liberated, and through the reducing effect of nascent hydrogen from the combination of sulphur in the food there are formed combinations of sulphur and the alkalies. These destroy the organic combinations of iron. In view of this, the later statement, that muriatic acid is a still more potent remedy for chlorosis than iron, becomes worthy of notice."

The sulphur of the sulphur-alkalies deprives the food in the intestinal canal of its iron, forming sulphuret of iron. This loss of iron causes a diminution of the material required for hæmoglobin.

If iron in large doses is ingested into

the stomach, and thence into the intestines, it combines with the sulphur in the sulphurets of the alkalies, but the iron in the food remains undisturbed by the sulphur. Thus it is that hæmoglobin may be formed in sufficient quantity.

A cure of chlorosis produced by iron is not a natural one; it is rarely permanent.

If the function of the stomach is improved by supplying muriatic acid, thus avoiding the formation of sulphurets of the alkalies, and the diminution in the formation of hæmoglobin, there will be a cure, but it will also lack permanence unless the epithelial cells of the peptonic glands which had lost the ability of forming muriatic acid, should regain it during this process.

If it is desired to supply the stomach in a natural manner with muriatic acid, we should, instead of muriatic acid, give *Natrum muriaticum* in a minimal dose.

This remedy will effect a permanent cure. (*Vide* the characteristics of common salt on page 58.)

That iron cannot cure chlorosis may appear from the fact that the serum of *venous* blood contains traces of iron, which is excreted by the kidneys in consequence of the retrogressive metamorphosis of the cells. The serum of *arterial* blood contains *no* iron. This proves that nature has no intention of patching blood-corpuscles by means of iron, or to influence them thereby in any way. Whoever in spite of this, endeavors to do so, does not act in agreement with nature. Chlorotic and anæmic patients must receive *new* blood-corpuscles in the manner indicated in the preceding article.

FACIAL DIAGNOSIS.

Two Spanish students wandering from Peñafiel to Salamanca discovered near the highway a tombstone on which were engraved the words: "*Aqui está enterrada el alma del licenciado Pedro Garcia.*" (Here is interred the soul of the licentiate Peter Garcia.) One of the two students laughed, because he did not understand the meaning and intention of the inscription. He went on. The second student remained behind; he lifted up the tombstone and found under it a purse containing gold coins; a note lying by it stated that these coins were intended for the person who would guess the meaning of the inscription.

My intention in calling up this old tale will be seen from what follows, as we shall treat of facial diagnosis, which will be judged of variously by the readers of these lines. He who only uses biochemical remedies, if he will practice his powers of observation, will in the course of time acquire the faculty of recognizing in many cases, especially in chronic diseases, from the physical state of the face and from its physical expression, which one of the bio chemical remedies will correspond with a given disease. Such a facial diagnosis ought not, indeed, of itself to determine the choice of the remedy to be used, but it may facilitate, respectively confirm the selection.

Whoever wishes to learn this facial diagnosis must acquire it in an autodidactic manner. The attempt to acquire it by means of a printed direction would lead to mistakes. A shepherd knows every individual member of his flock; but he will be unable to indicate the deciding characteristics.

Whoever would acquire facial diagnosis should give his particular attention first to *one* species of faces. The common-salt-face—if I may be allowed to compound such a word—is most easy to recognize. He should impress on his memory the quality and expression of the faces of those persons whom he has cured in a proportionally quick manner with *Natrum muriaticum.* A red thread will, as it were, run through the several impressions; he will recognize a family likeness.

Having first secured the common salt face, let him next pass to another soda face.

It is, of course, not necessary to state that physicians who are accustomed to give two or even more remedies in rapid alternation will never acquire facial diagnosis. Giving two remedies in alternation is permissible only very exceptionally, in cases where it, is or appears to be, unavoidable.

He who has once acquired this diagnosis will be convinced that it is just as important in a therapeutic sense as the soul of licenciate Peter Garcia was in a pecuniary sense.

Whoever may doubt the possibility of a facial diagnosis may be interested in the following case:

In the clinic of a university a man had died as to whom the clinical professor and his assistant physicians, in spite of their application of all diagnostic means, had not succeeded in making a diagnosis. When the body had been transferred to the professor of pathologic anatomy, he exclaimed as soon as he saw it: "*Cancer of the liver!*" and this diagnosis was verified by the dissection.

Of course no one can acquire facial diagnosis who, besides biochemistry, also uses all other kinds of curative methods, *e. g.*, if after giving a biochemical remedy he uses electricity or massage, or wrapping in wet sheets, or if he uses a so-called *Lebenswecker* (stimulator of life), pricking the skin of the patient and rubbing in so-called "*Mueckenfett*" (fly-fat). When a patient recovers by such a procedure no one can know to what to ascribe his cure. It may be, indeed, indifferent to the patient to what process he owes his cure, but this cannot be indifferent to the physician, for he has not learned anything from the case.

It would be a great mistake if anyone should expect to hasten the treatment of a biochemical case by various different remedies; the contrary would be the case in all probability. If we consider that the particles of the biochemical remedy cause molecular motions in the seat of disease which are to regulate the molecular motions, which have suffered pathogenic disturbances, it may be manifest that the molecular disturbances caused by electricity, massage, etc., must disturb the others just as, *e. g.*, the swinging of the pendulum regulating the mechanism of a clock would be disturbed if we should endeavor to hasten it by sudden impacts.

It has been asserted by certain persons that biochemistry will not suffice in all cases. I would request those who make these assertions carefully to study facial diagnosis. When you have mastered that, you may find a case where you will feel yourself called upon to use, *e. g.*, *Magnesia phosphorica* in a septic case. In consequence you will establish a cure. The difference between the remedy in question and *Kali phosphoricum* may not, however, in the meantime be exactly expressed in words.

EPILOGUE.

Those readers who have followed the development of my therapy from one edition to the other will remember that I have endeavored to remove mistakes made in the beginning and to insert new indications in my little work. This treatise was translated several years ago into English, into Spanish and into French. In these books, beside the errors already mentioned, there are many indications supplied by the translators which are either insignificant or erroneous.

When a translator, owing to his lack of apprehension of the subject, introduces his own *false* views into the translation, he injures the cause and discredits the author with his readers, who have no inkling of the fact that the translator has taken the liberty of adding the products of his own wisdom to the translation.

INDEX.

Printed in Great Britain
by Amazon